JN046194

The latest information in simple terms

Photocatalysis Experimental Methods

Akira Fujishima

Tsuyoshi Ochiai, Kengo Hamada, Donald Alexander Tryk,
Chiaki Terashima, Norihiro Suzuki, Katsunori Tsunoda,
Hitoshi Ishiguro, Jinfang Zhi, Jong-ho Kim,
František Peterka, Henrik Jensen,
and Photocatalyst Museum

Japan is leading the world in photocatalysis. Photocatalysis is playing an active role as an eco-friendly and clean technology to create a sustainable society. In particular, the strong oxidizing power of the titanium dioxide surface exposed to light has been applied to sterilization, deodorization, and antifouling. In addition, the surface of titanium dioxide becomes super hydrophilic under light irradiation, which prevents fogging of glass and increases the self-cleaning effect, and the field of application is expanding worldwide.

In particular, we have recently been in a difficult situation due to the novel coronavirus (COVID-19), and titanium dioxide, as a photocatalyst, is effective against the coronavirus, so it is attracting attention from various fields. For example, various products, including photocatalytic air purifiers, are now on the market and have been well received.

This book starts with the basic physical properties of titanium dioxide, and then explains the characteristics of titanium dioxide as a photocatalyst. Many specific coating materials are listed, and their coating methods and photocatalytic product evaluations according to JIS and ISO are summarized. In addition, specific product groups are categorized by function and explained. It also summarizes water splitting and future prospects of photocatalytic systems that use sunlight to decompose water and produce hydrogen.

In addition, in relation to the coronavirus, which is a recent central theme, specific experimental processes are explained in an easy-to-understand manner with photographs.

It also introduces the Photocatalyst Museum, where photocatalytic products are exhibited, the Photocatalysis Industry Association, where nearly 100 companies participate, and representative examples of photocatalytic products and research in China, Korea, and Europe.

We, the group of the Photocatalysis International Research Center at Tokyo University of Science and the researchers of the Photocatalysis Group (former KAST Group) of the Kanagawa Institute of Industrial Science and Technology (KISTEC), who have been studying photocatalysis for many years, have compiled this book focusing on experiments. We hope you will find this book useful, as it is compiled mainly from a wide variety of information sources available.

We hope that photocatalysis will be properly understood and that highly effective photocatalytic products will be widely used.

On behalf of the authors

Akira Fujishima

Contents

Chapter 7

Light Source Systems (Wavelength Characteristics, Intensity, Lifetime, Price, etc.) 095

Chapter 8

Apparatus System 111

Chapter 9

Product Examples 121

Chapter 10 Antibacterial and Antiviral Performance Evaluation Method

Chapter **1**

Basics of Photocatalysis
(Why Titanium Dioxide?)

Crystal structure of titanium dioxide and photocatalytic activity

There are three types of natural titanium dioxide crystals. They are called rutile, anatase and brookite, and have the same chemical formula of TiO_2, but they have different crystal structures, as shown in Figure 1-1.

The two main types of TiO_2 used in industry are rutile and anatase. The brookite type is of more academic interest.

Titanium dioxide of the rutile type is used as a white pigment and paint. On the other hand, the anatase type is mainly used as a photocatalyst.

Rutile titanium dioxide, which has relatively low photocatalytic activity, is used as a pigment for white paints and coatings, and is coated with silica to prevent the binder from being degraded. On the other hand, when photocatalytic activity is required, the anatase type is mainly used.

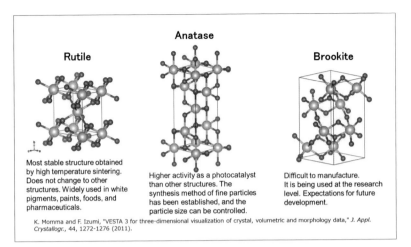

Rutile

Most stable structure obtained by high temperature sintering. Does not change to other structures. Widely used in white pigments, paints, foods, and pharmaceuticals.

Anatase

Higher activity as a photocatalyst than other structures. The synthesis method of fine particles has been established, and the particle size can be controlled.

Brookite

Difficult to manufacture. It is being used at the research level. Expectations for future development.

K. Momma and F. Izumi, "VESTA 3 for three-dimensional visualization of crystal, volumetric and morphology data," *J. Appl. Crystallogr.*, 44, 1272-1276 (2011).

Fig. 1-1 The three crystal structures of titanium dioxide.

Titanium dioxide is a kind of semiconductor

The word "semiconductor" means "half-conductor."

A conductor is a substance that conducts electricity. For example, copper and aluminum, which are widely used in power lines, are conductors. On the other hand, materials such as glass and rubber do not conduct electricity and are called insulators. Between conductors and insulators are semiconductors. Semiconductors are substances that can transmit electricity under certain conditions.

These conditions include the application of heat or light.

There is a wide range of materials that can be called semiconductors. Silicon (Si) and germanium (Ge) are so-called elemental or single-element semiconductors.

Semiconductors made from two or more elements, such as gallium arsenide (GaAs), are called compound semiconductors.

Titanium dioxide (TiO_2) and zinc oxide (ZnO) are also known as oxide semiconductors. Other types of semiconductor include amorphous semiconductors, which are used in solar cells.

Semiconductors can be divided into n-type and p-type, where the n in n-type stands for negative, meaning that the semiconductor contains impurities that produce electrically negative electrons (e^-). In the case of titanium dioxide, the oxygen in the crystal is removed and it acts like an impurity semiconductor, so it can be called an n-type semiconductor.

Characteristics of titanium dioxide

Oxide semiconductor

- It has semiconducting properties due to sites where oxygen is missing (defects) from the crystal structure.
- Impurities like sulfur can be added.
- Chemically stable and insoluble in acids and alkalis.
- It is earth-abundant.
- The color of the powder is white, but ultrafine particles (nanoparticles) are colorless when suspended with water or alcohol.
- Transparent thin films can be obtained by coating the ultrafine particles on a substrate and sintering.

Chapter 1-3

Semiconductor band structure and band gap energy

Band theory is used to explain the properties of semiconductors. We use the diagram shown in Figure 1-2. It is made up of a conduction band, in which freely moving electrons exist, and a valence band, in which holes, which are left over from electrons, exist.

The energy width of the forbidden band between the valence band and the conduction band is called the band gap energy.

In order to increase the number of electrons in a semiconductor, the electrons in the valence band must be brought up to the conduction band by applying energy greater than this band gap. This can be done by irradiating the semiconductor with light. In a semiconductor with a large band gap, the electrons in the valence band cannot normally be raised to the conduction band. However, when they receive external energy (for example, when irradiated with light corresponding to the band gap), the electrons in the valence band can be raised to the conduction band (excitation), leaving a number of vacancies (electron holes) in the valence band equal to the number of excited electrons. In the photocatalytic reaction, light is shone

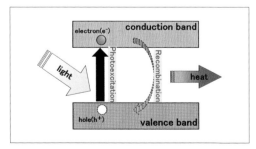

Fig. 1-2 Effect of light irradiation on semiconductors.

on the surface of titanium dioxide, and this action is illustrated in Figure 1-2.

The question is, what kind of light is effective? Light of a wavelength with a power greater than the band gap energy is required (see Figure 1-3).

Incidentally, the conversion of the band gap energy (E_g) to wavelength (nm) can be done by a very simple calculation: 1240 divided by E_g. Using this calculation, we can see that electrons in the valence band of rutile-type titanium dioxide ($E_g = 3.0$ eV) and anatase type titanium dioxide ($E_g = 3.2$ eV) can be raised to the conduction band by exposing them to light with wavelengths of about 413 nm and 388 nm, respectively.

In addition, as many holes are created in the valence band as there are electrons in the conduction band. Some of these holes and conduction electrons recombine to produce heat. Some of them also move to the surface and cause a surface reaction. This is the photocatalytic reaction.

There are three factors in the band structure of semiconductors that have the greatest influence on the photocatalytic reaction: (1) the band gap energy E_g (the energy difference between the lowest point of the conduction band and the highest point of the valence band), (2) the position of the lowest point of the conduction band, and (3) the position of the highest point of the valence band.

These factors vary greatly depending on the type of semiconductor. This is why each type of semiconductor has different physical properties (see page 13 for the band structures of the main semiconductors).

It is mainly the band gap energy that determines the wavelength of light that is effective in a photocatalytic reaction, and the position of the highest point of the valence band that determines the oxidative decomposition power of a photocatalyst.

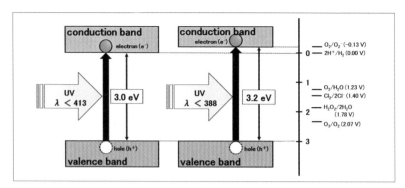

Fig. 1-3 Band gap of rutile (left) and anatase (right).

Titanium dioxide photocatalyst uses near ultraviolet light

The band structure of titanium dioxide determines the effective wavelength of light. The wavelengths of light are shown in Figures 1-4 and 1-5.

As mentioned above, titanium dioxide absorbs near-ultraviolet light of wavelengths below 400 nm, and the reaction proceeds. Then, what about the case of 254-nm light, which has higher energy, and is currently used in germicidal ultraviolet lamps?

In this case, 254-nm light is absorbed by the DNA of the constituent compounds in living organisms and causes damage to the DNA, such as the formation of dimers of pyrimidine (one of the bases that make up DNA).

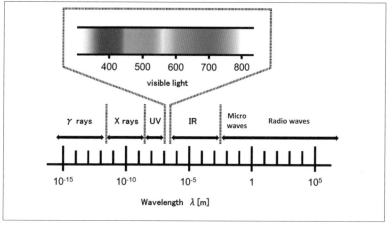

Fig. 1-4 Electromagnetic spectrum.

Titanium dioxide photocatalysts do not require ultraviolet light as short in wavelength as 254 nm, which is very energetic and harmful to the human body; light around 400 nm is found in sunlight, fluorescent lights, and recently, LEDs have been developed. Thus, titanium dioxide photocatalysts have excellent properties in that the reaction proceeds with relatively long-wavelength near-ultraviolet light.

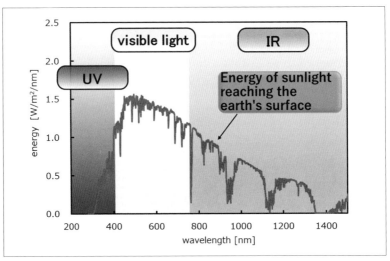

Fig. 1-5 Optical environment surrounding the earth.

Characteristics of titanium dioxide photocatalysts

· Absorb and work with light of wavelengths shorter than 400 nm.

· Absorption of sunlight, fluorescent light, and LEDs.

· Strong reactive power is generated on the surface.

   (decomposes almost all organic materials)

· However, they work only on objects in contact with the surface.

· They can decompose organic matter of an amount related to the number of photons absorbed (cannot decompose large amounts of matter)

· Only compounds adsorbed on the titanium dioxide surface can be decomposed.

Mechanism of photocatalytic oxidation and decomposition reaction

Photocatalytic reaction mechanism

The reaction mechanism of photocatalysis is still under investigation. In general, it is thought that various reactive oxygen species are generated by redox reactions on the surface, and these act as reaction intermediates to oxidize or reduce various compounds adsorbed on the surface (Fig. 1-6).

① $TiO_2 + h\nu \Rightarrow e^- + h^+$
② $e^- + O_2 \Rightarrow O_2^{\cdot -}$
③ $O_2^{\cdot -} + H^+ \Leftrightarrow HO_2^{\cdot}$
④ $h^+ + H_2O \Rightarrow {\cdot}OH + H^+$
⑤ $h^+ \rightarrow h_{trap}$

When titanium dioxide absorbs UV light, electrons (e^-) and holes (h^+) are generated inside the titanium dioxide (1). Of these electrons and holes, those that diffuse close to the surface are involved in the reaction. The electrons react with oxygen adsorbed on the surface to form superoxide anions ($O_2^{\cdot -}$) (2). In the presence of water, this $O_2^{\cdot -}$ is in equilibrium with the perhydroxyl radical ($HO_2^{\cdot}$), which is bound to a proton (H^+) (3). Holes, on the other hand, react with adsorbed water to produce hydroxyl radicals (4) or are trapped by surface atoms, i.e., surface trapped holes (h_{trap}) (5). The hydroxyl radicals produced are considered to be the main active species in photocatalytic oxidative decomposition because of their high oxidative activity.

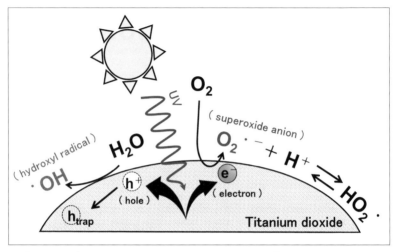

Fig. 1-6 Oxidation reaction mechanism of photocatalysis.

How does it inducer superhydrophilicity?

The phenomenon of "clouding" of glass is caused by the diffuse reflection of light by the numerous small water droplets on the surface. A titania surface that has been sufficiently exposed to light will not become cloudy. When the wettability of titanium dioxide is increased by light irradiation, water droplets spread over the surface and form a uniform film of water on the surface, thus preventing diffuse reflection (Fig. 1-7). This means that when water droplets come into contact with the titanium dioxide surface, the contact angle is reduced from the usual 70-80 degrees to almost zero, i.e., the surface becomes superhydrophilic. The surface of titanium dioxide before light irradiation is uniformly hydrophobic and has a large catalytic angle, but it becomes hydrophilic with light irradiation and finally becomes uniformly superhydrophilic. The reaction mechanism has been investigated and is illustrated in Fig. 1-8.

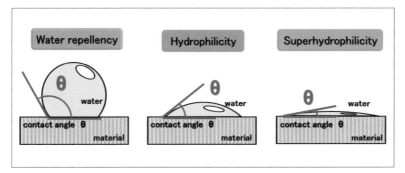

Fig. 1-7 Water repellency, hydrophilicity, and superhydrophilicity based on contact angle.

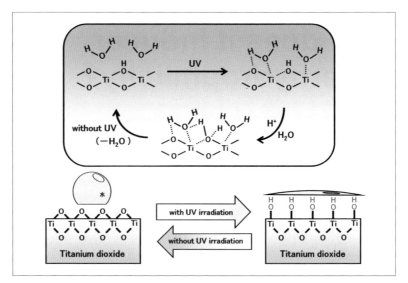

Fig. 1-8 Mechanism of superhydrophilicity.

What is superhydrophilicity?

· The contact angle of water droplets with the surface is as close to zero as possible.

· Water is present but forms a flat sheet on the surface.

· No fogging of glass or other surfaces.

· Oily substance can be removed from the surface when water is applied.

· Various applications are possible.

What about photocatalysts other than titanium dioxide?

We have mainly discussed titanium dioxide, but there are many other semiconductors with photocatalytic activity.

Nevertheless, titanium dioxide is the most widely studied and practically used.

Figure 1-9 shows the band structures of major semiconductors.

Many materials with a smaller band gap than that of titanium dioxide undergo autolysis when exposed to light in water. This is due to the fact that the holes generated by the light irradiation oxidize themselves, causing the metal ions to dissolve.

This is the reason why most semiconductors are not durable enough to be used as practical materials. In addition, there are resource and toxicity issues.

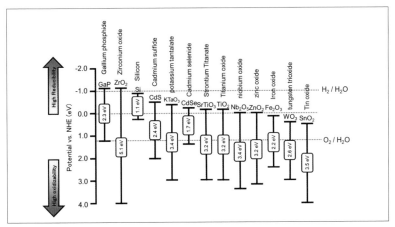

Fig. 1-9 Band gaps of various photocatalysts.

Photocatalytic systems other than titanium dioxide that are being studied.

· Oxides

$SrTiO_3$	Stable in water but have a large band gap ($SrTiO_3$) or a small
WO_3	reducing power (WO_3).
Fe_2O_3	
MoO_2	
etc.	

· Single-element semiconductors

Si	Can absorb visible light but are unstable in water.
Se	
Ge	

· Compound semiconductors

CdS	Can absorb visible light	
GaAs	but decomposes easily	
GaP	Toxic	disadvantages.
InP	Low in resources	

(Akira Fujishima)

Chapter **2**

Photocatalysis and Their Applications

Oxidative decomposition and superhydrophilicity and their applications

Six major functions of photocatalysis and their applications

Oxidative decomposition and superhydrophilicity and their applications

The main feature of titanium dioxide photocatalysis is the reaction that occurs on the surface of titanium dioxide when it is exposed to light. The photocatalytic reaction can be divided into two main categories. The first is the oxidative decomposition of oily organic matter, bacteria and viruses that come to the titanium dioxide surface. This reaction is powerful and can be used to break down any organic material into carbon dioxide and water. The other reaction is to change the properties of the titanium dioxide surface to superhydrophilic. Superhydrophilic is the opposite of hydrophobic, which means that water does not form water droplets on the surface, but spreads over the entire surface in a thin film. The whole concept of photocatalytic products is to create a clean environment by applying these two reactions.

In the industrial world, there were people who thought about applying this photocatalysis to various products, and this trend has grown to form various product groups, as we see today. These products are mainly those that clean themselves and those that clean the surrounding environment. Figure 2-1 summarizes the applications of photocatalysis. More specific application examples are given in Chapter 9.

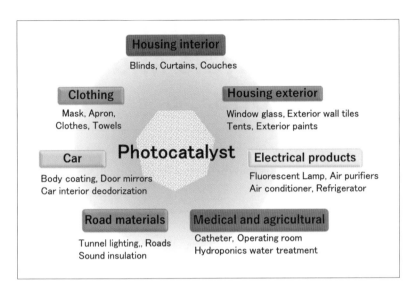

Fig. 2-1 Applications of photocatalysis.

Six major functions of photocatalysis and their applications

When a photocatalyst works, it can be classified into three functions, mainly as the property to clean the surface by photocatalysis.

① Anti-bacterial and anti-viral
② Anti-fouling
③ Anti-fogging

Of course, it is known that it is effective in reducing bacteria floating in the air.

In contrast, the properties that clean the surrounding environment are

④ Deodorization
⑤ Air purification
⑥ Water Purification

In contrast to this, there are three functions that help to clean the environment.

Deodorization and air pollution are the same in the sense that they both clean the air (air purification), but they target different substances and different locations, so they should be considered separately.

Figure 2-2 summarizes this.

In order for photocatalysis to occur, the titanium dioxide surface must be exposed to light. The light must be of a wavelength that titanium dioxide can absorb, and the intensity of the light must be carefully considered. Figure 2-3 summaries the light conditions in which photocatalysis operate.

Six major functions of photocatalysts

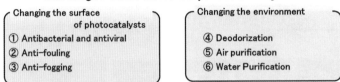

Changing the surface of photocatalysts
① Antibacterial and antiviral
② Anti-fouling
③ Anti-fogging

Changing the environment
④ Deodorization
⑤ Air purification
⑥ Water Purification

Fig. 2-2 Photocatalysis has two main roles.

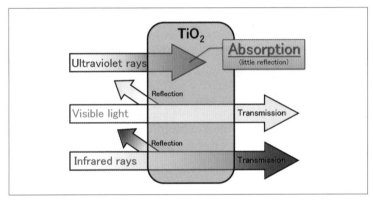

Fig. 2-3 Light that can be absorbed by titanium dioxide.

Photocatalysis, as an ecological and clean technology, is playing an active role in creating a sustainable society. (From the "Introduction" of this book)

Fig. 2-4 Sustainable development goals (SDGs).
Source : United Nations Development Programme HP

(Akira Fujishima)

Chapter **3**

Preparation of Coating Materials

Size and classification of titanium dioxide particles, how to make nanoparticles

Types of titanium dioxide coating agents

Preparation of titanium dioxide coatings

Chapter 3-1

Size and classification of titanium dioxide particles, how to make nanoparticles

Titanium dioxide has long been used as a pigment in paints, inks and resins and is produced in Japan at a level of hundreds of thousands of tons annually. The titanium dioxide that is used as a photocatalyst is one order of magnitude smaller in size than the titanium dioxide used industrially, which has a particle size of several hundred nanometers*. As shown in Fig. 1-4 in Chapter 1, light has the characteristics of a wave, and the length of the wave (wavelength) determines the characteristics of the light, such as ultraviolet or visible light. When light waves hit titanium dioxide particles, they are scattered in all directions (called

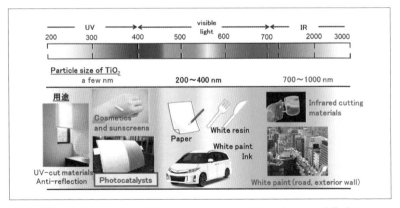

Fig. 3-1 Relationship between particle size and light wavelength, main applications.

scattering*). Large waves (waves with long wavelengths) are easily scattered by large particles, and small waves by small particles, so there are applications for each particle size of titanium dioxide (Figure 3-1).

*1 nm is equivalent to 0.000001 mm (millimeter), and 1 nm is equivalent to 10 Å (angstrom). The diameter of the novel coronavirus is about 100 nm. Nanotechnology is a term used to describe the area of science and technology that deals with materials on the size scale of nanometers.

*Light scattering: titanium dioxide particles of 200-400 nm in diameter appear white because they scatter all visible light. Clouds also appear white because they scatter all visible light, although the particle size varies greatly. Titanium dioxide particles with a diameter of 6 to 50 nm scatter blue visible light, so they look blue when dispersed in water. The reason why the sky looks blue is because the air molecules scatter blue light. Using this principle, it is possible to create a "blue sky and sunset" as shown in

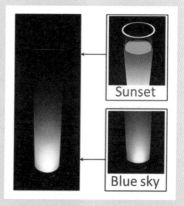

Sunset

Blue sky

the picture on the right. Here, titanium dioxide nanoparticles are dispersed in water and placed in a cylinder, which is illuminated from below by an electric lamp. The blue light contained in the light from the electric lamp is scattered at the bottom to create a 'blue sky', while the red light remaining at the top after the blue light has disappeared creates a 'sunset'. The same experiment can be carried out using milk, which is rich in protein.

In other words, the titanium dioxide that is used for photocatalysis is classified as "nano-sized titanium dioxide," and the manufacturing method is different from that of the titanium dioxide that is used for pigments. Figure 3-2 shows the difference in manufacturing methods between titanium dioxide for pigments and nano-sized titanium dioxide.

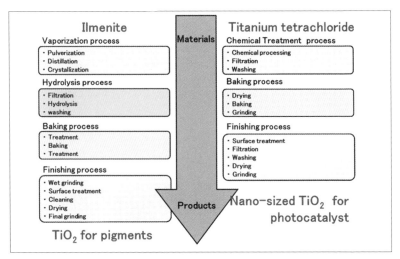

Fig. 3-2 Difference in manufacturing methods between titanium dioxide for pigments (right) and nano-sized titanium dioxide (left).

Types of titanium dioxide coating agents

Titanium dioxide for photocatalysis exists in the form of fine particles with a diameter of several nm to several tens of nm, and, if it is used as it is for water purification, the titanium dioxide particles must be removed by a filter after treatment. Therefore, from a practical point of view, it is necessary to adhere (i.e., coat) titanium dioxide particles to the surface of the substrate. In this case, there are two common methods (Fig. 3-3).

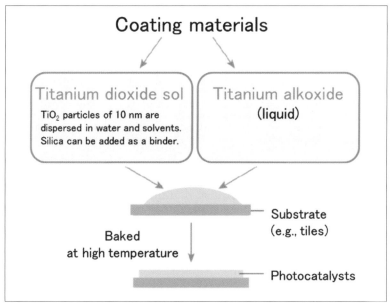

Fig. 3-3 Two types of coating methods.
Source: "All About Photocatalysis Revealed by Leading Experts", by Akira Fujishima (Diamond Inc., 2017).

The first is the sol-gel method, in which a substance is dispersed in a fluid state. Milk and mayonnaise are sols in which proteins and lipids are dispersed), the solvent part is removed to produce a non-flowable gel (a state in which a substance is dispersed without fluidity; tofu and konjac or konnyaku (Japanese yam cake) are sol-gels). Tofu and konnyaku are kinds of sols in which proteins and sugars are dispersed. Titanium dioxide sols are made by dispersing titanium dioxide particles as fine as 10 nm in water or alcohol, which, when coated on a substrate surface, eventually gel and form a transparent thin film. The second method is to use a substance (precursor) which becomes titanium dioxide by a chemical reaction. The precursor is a compound of titanium and alcohol (titanium alkoxide, e.g., $Ti(OCH_2CH_3)_4$). This is a compound of titanium and ethanol, called titanium tetraethoxide. When dissolved in an alcoholic solvent and coated, a glass-like titanium dioxide film is formed (see Chapter 4 for coating methods). Transparent crystalline titanium dioxide can be obtained by baking it at several hundred degrees Celsius. In any case, titanium dioxide sols and titanium alkoxides do not adhere to the substrate unless they are baked at high temperatures. Titanium dioxide coating solutions have been developed by adding a binder (adhesive) to titanium dioxide sols, which hardens near room temperature. Binders are often inorganic materials (e.g., silica) which do not decompose due to the photocatalytic reaction of titanium dioxide. Thus, a variety of coating solutions have been developed to fix titanium dioxide on the surface of various materials. In recent years, water-soluble peroxotitanic acid $(Ti(OOH)(OH)_3)$ has been used as a precursor, and the solvent used for the coating is water-based. This has the advantage that the coating can be applied to plastic substrates without damaging the substrate with organic solvents. The Table at the end of this chapter lists the coating solutions that are PIAJ certified by the Photocatalysis Industry Association (see Chapter 14).

Preparation of titanium dioxide coatings

The following is an example of preparation of titanium dioxide coating solution. (This is an experiment conducted by the author, referring to the description in "Photocatalytic Standard Research Methods" (Tokyo Book, 2005) by Bunsho Otani. Titanium dioxide thin film can be prepared by

1 As materials, 100 mL of 2-propanol, 1 g of titanium(IV) tetra-2-propoxide $(Ti(OPr^i)_4)$, and 4 g of titanium dioxide powder were prepared.

2 Put 50 mL of 2-propanol into a beaker, and while stirring with a magnetic stirrer, slowly drop 1 g of titanium(IV) tetra-2-propoxide into the beaker with a Pasteur pipette, and continue stirring overnight at room temperature.

3 After stirring overnight (the mixture becomes slightly cloudy), add another 50 mL of 2-propanol.

4 Add 4 g of titanium dioxide and stir for several hours to complete the reaction.

coating with this solution using a brush.

After coating, titanium(IV) tetra-2-propoxide reacts with moisture in the air or decomposes during heating to form a network of titanium dioxide. This network binds the titanium dioxide particles to each other and to the substrate.

In the case of soda-lime glass, the most commonly used glass material, the sodium in the glass diffuses during sintering and reacts with titanium dioxide to form sodium titanate (Na_2TiO_3), which has no photocatalytic properties.

To prevent the loss of photocatalytic properties, a layer of silica (SiO_2) is inserted between the glass substrate and the titanium dioxide layer. The dense silica layer then prevents the diffusion of sodium and protects the photocatalytic layer on the surface. The necessity of introducing silica into normal paint surfaces and the role of silica in the introduction of titanium dioxide into fibrous substrates are summarized in Figure 3-4.

A two-step coating method is also often used, in which an "undercoat" is applied to provide an intermediate silica layer, followed by a "topcoat" to provide a photocatalytic layer.

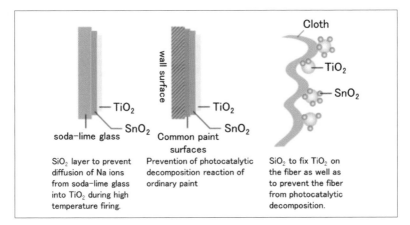

Fig. 3-4 Role of silica interlayer.

Source: "All About Photocatalysis Revealed by Leading Experts", by Akira Fujishima (Diamond Inc., 2017).

In addition, when photocatalyst is coated on polymer film such as PET, a hybrid polymer, which is a mixture of organic and inorganic materials at molecular level, is used as an adhesive layer in the middle to prevent the photocatalytic degradation of the film itself and to improve the adhesion between the film and the photocatalyst layer (Fig. 3-5). A film formation technique has been developed that allows the composition of the film to change continuously from organic to inorganic in a layer of approximately 100 nm, with only organic components on the surface in contact with the organic polymer film and only inorganic components on the surface in contact with the titanium dioxide. Thus, the latest developments in materials science at the nanometer level are strongly encouraging the industrialization of photocatalysis.

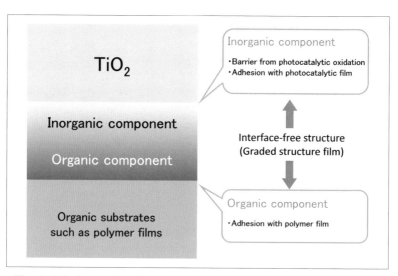

Fig. 3-5 Schematic of cross section of graded structure film.

Source: "All About Photocatalysis Revealed by Leading Experts", by Akira Fujishima (Diamond Inc., 2017).

(Tsuyoshi Ochiai)

Types of coating methods

As shown in Fig. 4-1, there are two main methods for coating photocatalysts on substrates.

One is the "wet process method", in which a photocatalytic coating liquid is directly coated on a base material and includes impregnation, brushing, spray coating, roll coating, spin coating and dip coating. These methods require only little initial investment cost of equipment and are relatively easy to apply. Therefore, they are used in a wide range of applications, from substrates of a few centimeters in size to the actual buildings.

The other method is the "dry process method", where a base material is coated with photocatalyst mainly in a vacuum, includes sputtering, vacuum evaporation, ion plating and CVD (Chemical Vapor Deposition). Although these methods require much higher initial investment cost, they can fabricate a dense, hard coating layer.

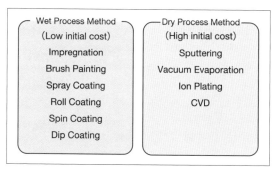

Fig. 4-1 List of coating methods.

Impregnation method

In the impregnation method, the coating is applied by immersing a base material in the coating liquid and then pulling it out to dry. Although this is a very simple method, a homogeneous film can be formed with good reproducibility with a precise withdrawing process (for details, see Chapter 4-7 Dip-Coating method).

If the base material to be coated is rod-shaped, the "squeeze coating" method is used, in which excess paint is rubbed off with a rubber or sealing material. As shown in Fig. 4-2, there are two ways of coating: by moving the base material and by moving the coating liquid. For a base material with a constant shape, such as bars, it is possible to coat the entire surface evenly in one pass. However, because the amount of coating liquid that is paintable at one time is not enough, several coats are sometimes necessary to increase the coated layer thickness.

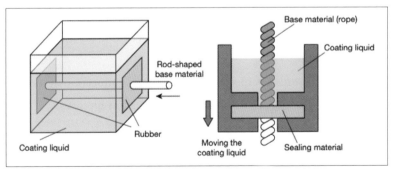

Fig. 4-2 Schematic diagram of squeeze-coating method.

Brush painting method

The brush painting method is a traditional coating method, which can be used on a wide range of shapes. It is an ecological way of using the coating liquid and is suitable for slow-drying coatings. The brush painting method consists of four stages.

1 Soaking

Soak around 2/3 of the brush with the coating liquid and absorb it. Tap the tip of the brush gently to prevent dripping.

2 Coating

Coat the coating layer by using the tip of the brush. For thin coating, the brush should be positioned almost perpendicular to the base material. The coating should be done from one side to the other side for a horizontal plane, while it should be done from bottom to top for a vertical plane. If the area is large, divide it into sections (approx. 80 cm square) and coat each section separately. If the surface has a long side and a short side, coat along the longer side.

3 Smoothing

Move the brush perpendicular to the coating direction of the first coating layer to make the coating layer thickness uniform.

4 Brushing

This process is used to ensure a uniform thickness and to prepare the brush marks. A fine-tipped brush is moved on the coated layer from corner to corner in a parallel direction.

Although brush painting is a simple method, it requires a lot of skill to achieve a beautiful finish.

Spray-coating method

In the spray-coating method, the coating liquid is micronized (like a mist) and sprayed on the base martial by means of compressed air. This method is very efficient and enables one to coat uniformly even over a large area. In addition, it can produce a beautiful coating on substrates whose shape is challenging for the brush painting method.

The spray-coating machine consists of an air spray gun and an air compressor. The air spray gun is a mechanical tool that mixes compressed air with the coating liquid. Spray guns can be divided into two types depending on the place where the coating liquid and air mix (Fig. 4-3). In an internal mixing gun, the air shears the coating liquid into strands and carries them to the base material. Because the internal mixing gun gives a rough finish coating, although it is used for ricin and tile finishes on exterior walls, it is not suitable for photocatalytic coatings.

On the other hand, external mixing guns, in which the coating liquid

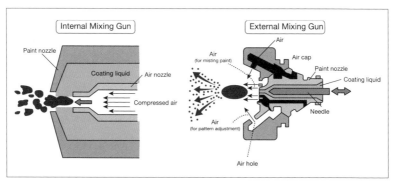

Fig. 4-3 Schematic diagrams of two types of air spray guns.

and the air contact outside the device, can micronize the coating liquid. The size of the micronized particles can be freely adjusted by the incoming air pressure, and the smaller coating liquid particles brings a higher specular glossiness to the coated surface. However, because the smaller particles tend to bounce, the air pressure should be set so that the micronized particles are within a certain size range.

This method is suitable for coating large areas such as the walls of existing buildings. However, the smoothness and transparency of the coated photocatalyst layer depends on the skill of the operator. Therefore, a certain amount of experience is required to achieve a high quality photocatalyst layer with good reproducibility. Other drawbacks are that the coating liquid loss is large, and masking is required to prevent undesired coating.

Roll-coating method

The roll-coating method uses a set of rollers to produce a uniform film of the coating liquid and then transfers it to the substrate. As shown in Fig. 4-4, the coating liquid is rolled up evenly on a pick-up roll and then transcribed to a doctor roll, which adjusts the film thickness. While maintaining the film thickness, the coating liquid is then transcribed to the coating roll and finally transcribed to the base material.

There are two types of roll coater, the "natural" type, in which the coating roll and the coated material are moving in the same direction, and the "reverse" type, in which the coating roll and the coated material are moving in opposite directions (Fig. 4-4). In the natural type, the coating liquid is spread over a base material by shearing the cross-section of the liquid film homogenized by the doctor roll. Therefore, rolling

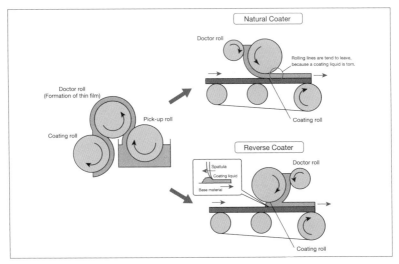

Fig. 4-4 Schematic diagrams of roll coaters.

lines tend to remain, and adjustment of the film thickness is difficult. On the other hand, in the reverse type, the spatula and the coated material are moving in opposite directions, which prevents rolling lines and ensures a uniform film thickness.

Spin-coating method

In the spin-coating method, the coating liquid is dropped on a plate and then the plate rotates at high speed to produce a thin film using centrifugal force (Fig. 4-5). Since the coating liquid has to wet the surface of the plate, it is important that the plate surface has good wettability with the coating liquid.

As the coating liquid spreads by centrifugal force and the viscosity increases by the solvent volatilization, the coating liquid becomes a gel (no longer fluid) and forms a thin film. Assuming that the centrifugal and viscous forces applied to the coating liquid on a rotating plate are balanced, and calculating the time-dependence of film thickness from the estimated velocity and flow rate of the centrifugal force direction from Newton's viscosity low, the theoretical film thickness (h) is expressed as follows

$$h = \frac{h_0}{\sqrt{1 + \frac{4\omega_0 h_0^2}{3v}t}}$$

where h_0 is the initial film thickness, ω_0 is the rotation rate, t is the rotational time, and v is the kinematic viscosity expressed as

$$v = \frac{\mu}{\rho}$$

where μ and ρ are viscosity and density, respectively.

The above formula clarifies that the film becomes thicker with increasing viscosity of the coating liquid, while it becomes thinner with increasing rotation rate. Thinner films will be obtained with increasing rotation time; however, the film thickness does not continuously decrease, because it converges, depending on the rotation rate.

Although the spin-coating method is a simple and relatively reproducible method of coating, the scattering of excess coating liquid leads to large losses of coating material, which is a significant disadvantage.

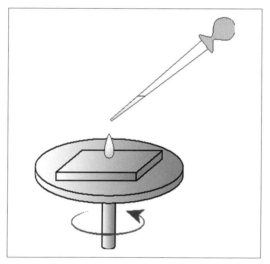

Fig. 4-5 Schematic diagram of spin-coating.

Dip-coating method

In the dip-coating method, a base material is dipped into the coating liquid and then withdrawn at a constant speed to produce a thin film (Fig. 4-6). Similar to the spin-coating method, the wettability between coating liquid and the material's surface becomes important.

When a base material is withdrawn from the coating liquid, it is attracted to the material's surface, where the solvent volatilizes and the thin film forms after gelation. Therefore, the film thickness is determined by the viscosity of the coating liquid and the withdrawing speed.

If both the withdrawing speed and the viscosity are high, the thickness (h) is determined by the balance between viscous resistance and gravity and is expressed as follows.

$$h = C\sqrt{\frac{\eta\,U}{\rho g}}$$

On the other hand, if the withdrawing speed is sufficiently slow and the viscosity is low, the surface tension at the gas-liquid interface dominates the film thickness and the following equation is established

$$h = \frac{0.94 \times (\eta\,U)^{2/3}}{\gamma^{1/6}\sqrt{\rho g}}$$

where C is the coefficient, U is the withdrawing speed, η is the kinematic viscosity, ρ is the density of the coating liquid and γ is the surface tension (0.8 for the Newtonian fluid used in dip coating). From these equations, it can be seen that the higher the viscosity of the coating fluid and the higher the withdrawing speed, the thicker the coating will be.

Because a higher viscosity in the coating liquid is required for a thicker film, thickening agents are sometimes added to the coating liquid. Most thickeners are organic compounds with a relatively high molecular weight and a high boiling point. However, because the solvent is less likely to evaporate with increasing coating liquid viscosity, the amount of solvent (and additives such as thickeners) remaining in the gel also increases. This sometimes causes cracking and poor film uniformity during the calcination (thermal treatment) for removal of solvent and organic compounds. To avoid this, it is necessary to build up thicker layers by repeatedly producing thin films using low-viscosity coating liquids.

When the base material to be coated is a plate, a thin film is formed on both sides. If a coating only on one side is demanded, it is necessary to wipe off the gel from the undesired side after the coating, or to tape the unwanted side in advance and remove it after the coating. The coating liquid tends to accumulate at the lower edge, resulting in an uneven coating. If necessary, the gel adhering at the lower edge should be removed.

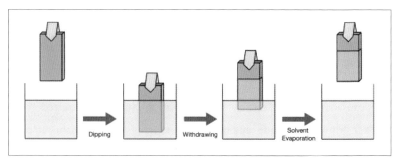

Fig. 4-6 Schematic diagram of dip-coating.

Chapter 4-8

Sputtering method

Sputtering (sputter deposition) is one of the most common methods of dry-process deposition. The figure below shows a typical sputtering method with a diode sputtering system. The basic principle can be explained from this diode sputtering method. As shown in Fig. 4-7, an inert gas such as argon is introduced into a vacuum chamber , and a high voltage is applied between the target (material to be deposited) and the substrate (surface to be deposited) to generate an internal glow discharge, which turns the inert gas into a plasma (ionization). The ionized inert gas is attracted at high speed to the target, which has a negative potential, and collides with the target material. The particles of the target material are ejected by the force of this collision and adhere to the surface of the substrate. By repeating this process, the target material at the bottom is deposited on the substrate at the top in a stable and dense state, without any change in composition, forming a thin film.

In addition to the diode sputtering method shown in the figure below, there are also magnetron sputtering methods that improve the deposition speed, and RF sputtering methods that can be used to deposit ceramics and other materials. There is also a reactive sputtering method in which clusters of metal oxides or metal nitrides are deposited on

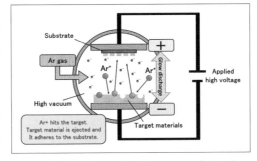

Fig. 4-7 Schematic diagram of diode sputtering method.

the substrate by introducing reactive gases (oxygen, nitrogen, etc.) in addition to the inert gas.

Vacuum evaporation

Vacuum evaporation is a method of forming a thin film by heating the material to be deposited, such as a metal or oxide, in a vacuum, causing it to evaporate or sublime, and then depositing the evaporated or sublimed particles on the surface of the substrate (see Figure 4-8). This is just an image, but it is easy to understand if you imagine the phenomenon that when a lid is placed on top of a pot of boiling water, steam adheres to the inside of the lid and forms water droplets. This is exactly the same principle as in evaporation coating. In vacuum evaporation, metals and oxides, which are difficult to evaporate at atmospheric pressure, are placed in a high vacuum environment to reduce their vapor pressure and vaporize them at a temperature of around $1,000°$ C. (This is the same as water boiling at less than $100°$ C at low atmospheric pressure.) In vacuum evaporation, the materials can be heated by resistance heating, high-frequency induction heating or electron beam heating. In resistance heating, an electric current is passed through a resistive element such as tungsten or molybdenum and the heat generated by the resistance is used as the heat source. It is suitable for low melting point metals such as gold and silver. High-frequency induction heating is

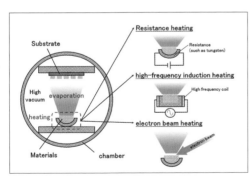

Fig. 4-8 Schematic diagram of vacuum evaporation method.

based on the principle of induction heating, where a conductor is inserted into a coil connected to an AC power supply and the conductor itself generates heat. The crucible in which the material is placed need only be a conductor, so carbon or similar material is used for the crucible. Finally, electron beam heating is a method of forming a thin film by irradiating a material with an electron beam, which heats and evaporates the material, as shown in the diagram. It is suitable for materials with a high melting point, such as oxides.

Ion-plating method

Ion-plating is very similar to vacuum evaporation. It is similar in that the material is vaporized by heating, but the difference is that the evaporated particles are accelerated in the direction of the substrate. In ion-plating, the evaporated particles pass through the plasma and become positively charged, thus accelerating them towards the substrate, which has a negative potential. The accelerated particles adhere to the substrate and are deposited, forming a thin film. Due to the acceleration of the particles, a denser and better adherent thin film can be produced than by vacuum evaporation (Fig. 4-9).

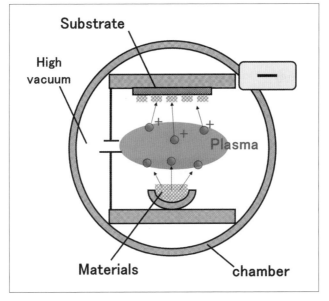

Fig. 4-9 Schematic of ion-plating method.

CVD method

CIn CVD (Chemical Vapor Deposition), gaseous raw materials are fed into a chamber at atmospheric pressure to medium vacuum, and energy such as heat, light or plasma is applied to cause a chemical reaction to synthesize the target substance and form a thin film on the surface of the substrate. In the case of metal oxides, a gas of a compound of the metal component of the target metal oxide is circulated over the substrate and the metal oxide produced by the chemical reaction is deposited on the substrate. The advantages of this process are that it does not require a high vacuum and therefore does not require expensive equipment, it is fast, and the composition and thickness of the film can be easily controlled (Fig. 4-10).

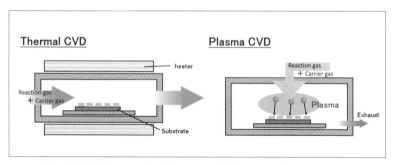

Fig. 4-10 Schematic diagram of plasma CVD method.

References
1. Minoru Tsubota, " わかる！ つかえる！ 工業塗装入門〈基礎知識〉〈段取り〉〈実作業〉(in Japanese)" (Nikkan Kogyo Shimbun, 2019)
2. " ドライプロセスによる表面処理・薄膜形成の基礎 (Introduction to Dry Processing for Surface Finishing and Thin Film Coating) (in Japanese)" edited by The Surface

Finishing Society of Japan (Corona, 2013)

(Norihiro Suzuki, Kengo Hamada)

Chapter **5**

Evaluation of Materials

Characterization of photocatalyst powder
Characterization of photocatalytic thin film

Characterization of photocatalyst powder

1　Particle size distribution

The particle size distribution (i.e., the proportion of particles of a given size) is measured by laser diffraction and scattering. When a particle is irradiated with a laser beam, it emits light (called diffracted and scattered light) in various directions, i.e., back and forth, up and down, left and right. The intensity of the diffracted/scattered light forms a spatial pattern (light intensity distribution pattern) in the light-emission direction, and the pattern varies, depending on the size of the particles. Therefore, it is possible to determine the particle size distribution by irradiating a group of particles with a laser beam and analyzing the light intensity distribution pattern emitted from the particles.

2　Specific surface area and pore volume

The gas adsorption method is used to calculate the specific surface area and pore volume distribution. They are determined from the adsorbed gas volume and the gas condensation of an inert gas molecule with a known adsorption cross-sectional area. Generally, nitrogen gas is used as the adsorbent gas, and the measurement is conducted at liquid nitrogen temperature (77 K).

As the amount of gas increases, the surface of the sample is covered with gas molecules, and after the entire surface is covered with gas molecules, new gas molecules are layered on top of the adsorbed gas molecules, resulting in multilayer adsorption. As shown in Fig. 5-1, this phenomenon appears as a change in the amount of adsorbed gas molecules with respect to pressure (i.e., adsorption isotherm). By applying the BET equation in the relative pressure region between the first layer adsorption and the multi-layered adsorption, the amount of adsorbed gas in a monolayer can be calculated accurately. The surface area of the sample can be calculated by multiplying the amount of adsorbed gas in a monolayer by the cross-sectional area of a single adsorbed gas molecule.

As more and more gas molecules are adsorbed onto the surface of the sample, they begin to condense in the pores (i.e., capillary condensation). At this point, a large number of gas molecules are transformed to a liquid state, and a rapid increase in the amount of adsorbed gas molecules can be observed on the adsorption isotherm. The pressure at which the condensation occurs is known to be correlated with the pore size. The increase of the adsorbed gas amount is proportional to the internal volume of the pores, and therefore, the volume distribution of the pores (i.e., pore size distribution) can be determined from the adsorption isotherm.

If the powder particles have pores, the adsorption and desorption processes do not coincide, and hysteresis is observed in the adsorption/desorption isotherm. This phenomenon is considered to be closely related to capillary condensation.

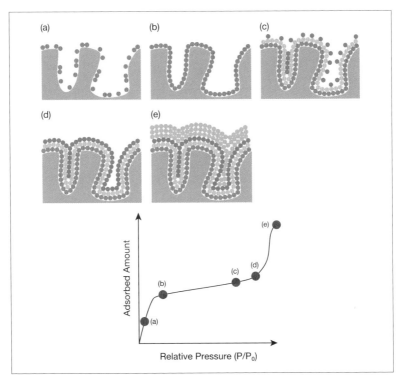

Fig. 5-1 Adsorption isotherm and schematic diagram of gas adsorption at each point.

3　Crystal structure

The crystal structure can be studied using X-ray powder diffraction (XRD), which is based on the analysis of the scattered and diffracted X-ray light produced when a sample is irradiated with X-rays (Fig. 5-2). When a sample is irradiated with X-rays, the lattice crystal of the material acts as a diffraction grating for the X-rays, and the X-rays are scattered and interfered with by the electrons around the atoms. This interference increases the intensity in a particular direction and gives a spectrum that reflects the crystal structure. The scattered X-rays intensify each other when $2d\sin\theta = n\lambda$ (Bragg's equation). where d is the lattice constant, θ is the angle of incidence of the X-rays, n is an integer and λ is the wavelength of the X-rays. Therefore, the spectrum of the X-ray diffraction can be obtained by irradiating a sample with incident X-rays of known wavelength and scanning a detector placed on the diffracted light side to measure the diffraction angle 2θ and the intensity of the diffracted X-rays at that time. The lattice parameter can be determined from the Bragg equation, the crystal structure from the spectral pattern, and the strain from the change in the crystal face spacing (d-value). In this way, XRD can provide information that reflects the crystal structure of the sample.

In the field of photocatalysis, XRD is also suitable for characterizing the crystal structure of a sample. Photocatalysts, such as titanium dioxide, are ionic crystalline materials, and XRD can provide the kind of information

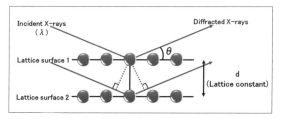

Fig. 5-2 Principles of X-ray diffraction.

described above. Titanium dioxide has three different crystal structures: anatase, rutile and brookite. Depending on its crystal structure, the degree of photocatalytic activity, such as organic decomposition performance, varies. Therefore, the crystal structure of titanium dioxide as a photocatalyst is very important.

4 Bonding state

Raman spectroscopy and X-ray photoelectron spectroscopy are used to evaluate the bonding state of a sample. Raman spectroscopy is an analytical method to evaluate the bonding state and crystallinity of a material by using the phenomenon that the frequency of the light incident on the material changes as it interacts with the material. A similar analytical technique is Fourier transform infrared spectroscopy (FTIR). The main difference between the two is the type of light being measured: FTIR measures the decrease in incident light, i.e., the absorption of light, whereas Raman spectroscopy measures the scattered light (at a different wavelength to the incident light) caused by the incident light. When light is incident on a material, it causes scattering of the same wavelength as the incident light (Rayleigh scattering). At the same time, there is a small amount of scattering at a wavelength different from that of the incident light (Raman scattering). The spectrum of the Raman scattering light reflects the crystal structure of the material, so that the bonding state and the crystallinity of the material can be evaluated from the information of the spectrum (Fig. 5-3).

The Raman spectrometer consists of a light source, a spectroscope and a detector. The signal obtained is a Raman spectrum, where the horizontal axis is the wavelength (wavenumber) and the vertical axis is the intensity. In general, the Raman spectrum is expressed as a shift from the wavenumber of the incident light, using the unit wavenumber [cm^{-1}], which is the reciprocal of the wavelength. Crystallinity can be determined from the peak width, and physical properties such as bonding state and stress can be evaluated from the amount of peak shift. Measurements can be made on samples in all states - solid, liquid and gas. However, materials that emit fluorescence can interfere with the

Raman spectra, making them difficult to measure.

In the field of photocatalysis, Raman spectroscopy can also be used to characterize the crystallinity and bonding state of a sample of a photocatalytic material (e.g., titanium dioxide). For example, the Raman spectrum of titanium dioxide doped with other elements shows a peak shift compared to that of titanium dioxide without doping, due to the change in the bonding state of the titanium dioxide crystal caused by the doped elements. In addition, if there is a change in the crystallinity of the titanium dioxide crystal, the peak width will also change.

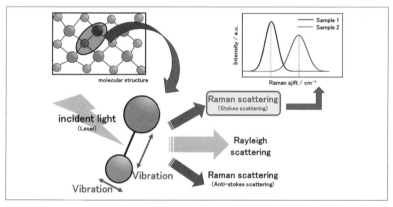

Fig. 5-3 Schematic diagram of Raman spectroscopy.

X-ray photoelectron spectroscopy (XPS) can be used to characterize the constituent elements and their chemical bonding on the surface of a solid sample. XPS uses X-rays as the excitation source, whereas the use of ultraviolet light results in a technique that is called ultraviolet photoelectron spectroscopy (UPS). UPS provides information on the semiconductor properties of solid materials, such as band gap and work function.

When a solid sample surface is irradiated with X-rays, photoelectrons are emitted from atoms excited by the X-rays (Fig. 5-4). The energy E of the photoelectrons emitted from the sample surface can be expressed by $E = h\nu - E_B - \phi$. Since $h\nu$ and ϕ are known, the binding energy E_B can be determined by observing the energy E of the photoelectrons. Since this binding energy is specific to each element, the position of the peak in the photoelectron spectrum can be used to identify the elements that make up the surface of the sample. Different atoms have different binding energies in different compounds (bonds) due to their interactions with surrounding atoms. These differences in chemical bonding cause the detected peak to vary by a few eV from the peak position of the single element (peak shift). The magnitude of this peak shift can be used to determine the chemical bonding state of the element of interest.

A unique feature of XPS is the ability to obtain information about the surface of the sample. The X-rays irradiated on the sample penetrate deep into the sample, and photoelectrons are generated even inside the sample. However, the photoelectrons produced have a maximum energy of only about 1,500 eV. The mean free path of these photoelectrons is so small that they are unable to reach the surface. Therefore, the spectra obtained from XPS are limited to the surface of the sample (down to a few nm in depth). XPS can be accompanied by a sputter etching mechanism using Ar^+ ions. This allows the depth profile of the material to be analyzed by repeated etching of the sample surface.

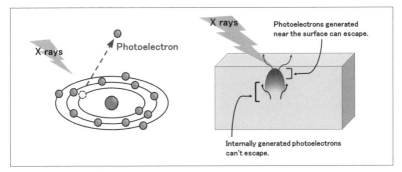

Fig. 5-4 Principle of X-ray photoelectron spectroscopy and the escape depth of photoelectrons.

5   Optical response and band gap

Ultraviolet-visible spectrophotometers (UV-VIS) not only measure the transmittance of light incident on a thin film or other material but also, by using an integrating sphere, the reflectance of a solid material at each wavelength. When light is incident on a powder sample, some of it is reflected back onto the surface of the powder. The rest of the light enters the powder, is reflected and refracted by the particles, and travels in various directions. If the light is of a wavelength that is not absorbed by the powder, it is reflected and refracted repeatedly before finally being emitted into the air. On the other hand, if the powder absorbs the light, it is gradually weakened by the repeated reflection and refraction processes. As a result, a diffuse reflection spectrum similar to the transmission spectrum is obtained.

Figure 5-5 shows the results of diffuse reflectance measurements of photocatalytic powders with different compositions. The yellow powder is a visible light-responsive photocatalyst, and the white powder is titanium dioxide. The figure below shows that titanium dioxide reflects almost 100% of the visible light with wavelengths longer than 400 nm. In other words, this material absorbs only ultraviolet light. In contrast, visible-light-responsive photocatalysts have low reflectance, especially in the 400-nm – 500-nm region, compared to titanium dioxide. In other words, this material absorbs visible light in the region of 400 nm to 500 nm. The results of the diffuse reflectance measurements can be transformed into a graph, as shown in Figs. 5-6 by performing a calculation known as the Kubelka-Munk transformation. The band gap of the material (the energy of light required for the photocatalytic reaction to occur) can be calculated from the rising point of this graph.

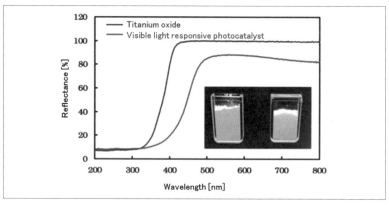

Fig. 5-5 Diffuse reflectance measurement of photocatalytic materials.

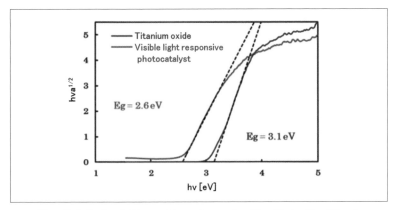

Fig. 5-6 Calculation of band gap by the Kubelka-Munk transform.

Sorce: Kanagawa Institute of Industrial Science and Technology HP
https://www.kistec.jp/sup_prod_devp/eval_devl/yuukip/hikarikeisokukiki/
sigaikashi/kk_040023_uv-vis_jirei 4 /

6　Sample morphology

The electron microscope is a powerful tool for observing the morphology of a sample, especially the microstructure at the nanometer scale. The Scanning Electron Microscope (SEM) allows us to observe the surface morphology of a sample at low and high magnification.

When a narrow beam of electrons is focused on a sample, secondary electrons, reflected electrons and characteristic X-rays are emitted from the surface of the sample. The amount of secondary electrons captured by the detector gives a two-dimensional image. The SEM measures secondary electrons, so there is no contrast between the images obtained for the same material, but, in reality, the amount of secondary electrons produced depends on the surface texture of the sample, so that, even for the same material, the image can reflect the surface texture. By measuring the characteristic X-rays emitted, qualitative analysis can be carried out. Qualitative analysis using characteristic X-rays requires the use of an SEM equipped with an Energy Dispersive X-ray spectrometer (EDX).

Fig. 5-7 shows FE-SEM images of the surfaces of photocatalytic materials with different geometries, where FE stands for field emission, and where the electron beam is narrowed down to a higher magnification than in conventional SEMs. The FE-SEM (field emission scanning electron microscope) is the most common type of SEM today. The images show that the surfaces of the two materials can be observed at very high magnification.

Transmission electron microscopes (TEMs) are also used to observe microstructures in the same way as SEMs: while SEMs produce images by detecting secondary electrons emitted from the surface of a sample, TEMs use a beam of electrons that passes through a sample that is thinned by processing it to a thickness of less than 100 nm. In contrast to the SEM, which produces images by detecting secondary electrons emitted from the surface of a sample, the TEM produces images by passing an electron beam through a sample that has been thinned to less than 100 nm in thickness and detecting the transmitted electrons. Depending on the structure of the sample and its constituent components, the electron beam penetrates the sample differently, resulting in the image shown in Fig. 5-8. Whereas the SEM is suitable for observing the surface of a sample over a reasonably large area, the TEM is more suitable for observing internal structures such as cross-sections over a much smaller scale.

Fig. 5-8 shows a TEM image of Au-loaded titanium dioxide (TiO_2), showing that the TiO_2 particles are 10-20 nm in diameter and are surrounded by Au particles of about 1 nm in diameter.

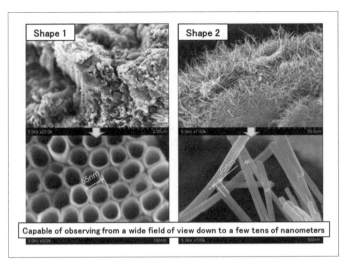

Fig. 5-7 Surface observation of photocatalytic materials by SEM.

Source: Kanagawa Institute of Industrial Science and Technology HP
https://www.kistec.jp/sup_prod_devp/test_and_mes/koudo/0300_bunseki_jirei/bj_fe-sem_01/

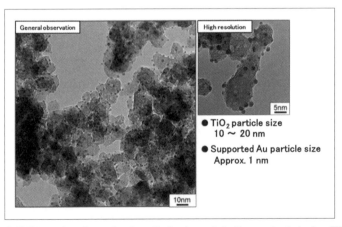

Fig. 5-8 Structural analysis of photocatalytic materials by TEM.

Source: Kanagawa Institute of Industrial Science and Technology HP
https://www.kistec.jp/sup_prod_devp/test_and_mes/koudo/0300_bunseki_jirei/bj_fe-tem_03/

Characterization of photocatalytic thin film

1 Adhesion of the thin film to the substrate

The adhesion of the thin film to the substrate (a base material) can be evaluated with the ultrathin film scratch tester, and the test method is standardized in Japan Industrial Standard (JIS R 3225). In this tester, a hard needle (diamond indenter) with a certain radius of curvature is pressed against the surface of the film and scratches the surface of the film while increasing the load, and the adhesion is evaluated by measuring the load value at which peeling of the film occurs.

A schematic of the detection unit of the ultrathin film scratch tester is shown in Fig. 5-9. A diamond indenter is attached to the tip of a cantilever extending from the cartridge. The indenter is pressed against the sample, while the cartridge is slightly vibrated horizontally toward the scratch direction. The frictional forces between the sample surface and the indenter cause the indenter to lag behind the horizontal movement of the cartridge. As a result, the relative position of the magnet attached to the cantilever and the coil in the cartridge changes, and an electrical signal is generated. At the same time, the indenter scratches the surface of the sample, while increasing the load applied to the sample by the indenter. When the film peels off, the electrical signal from the cartridge changes due to the unevenness of the sample surface and the change in the coefficient of friction, thus allowing the detection of film peeling.

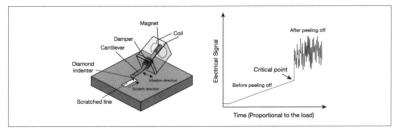

Fig. 5-9 Schematic diagram of the scratch testing machine (detection section) and detection signals.

2 Grain size (surface roughness) and physical properties on the surface

For the measurement of surface grain size (surface roughness), the probe type step profiler (stylus profiler) has been used for many years. A diamond-tipped stylus is used to trace the surface of a sample at a constant low pressure to measure steps, surface roughness and waviness. This method is highly reliable, because the sample surface is traced by physical contact; however, it is impossible to measure sticky or soft samples. Furthermore, the measurement of structure finer than the diameter of the stylus is also impossible.

Microscopic observation is the most common method of measuring without touching the sample. An optical microscope is sufficient for coarse samples with large irregularities, while a laser microscope (micrometer-level observation) or a scanning electron microscope (SEM) (nanometer-level observation) is required for smaller irregularities.

Different types of scanning probe microscope (SPM) allow atomic-level measurements to be made. In this type of microscope, a pointed probe is moved across the surface of the sample and the interaction between the probe and the sample is visualized (Fig. 5-10). It is now an indispensable tool in the study of surface properties, as it allows not only surface observation but also the imaging of various properties of the sample surface.

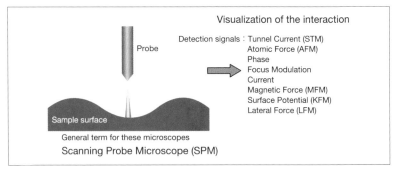

Fig. 5-10 Schematic diagram of types of scanning probe microscope.

3 Transparency

The transparency of a thin film deposited on a transparent substrate such as glass can be evaluated by the transmittance (T), which is the ratio of the transmitted light I to the incident light I_0.

$$T = \frac{I}{I_0}$$

Using a UV-Visible-Infrared spectrophotometer, the transmittance of light at each wavelength can be measured. Higher transmittance of visible light results in a more transparent appearance.

Even though the light is transmitted, if light diffusion occurs, the transparency decreases. The transparency of a specimen is evaluated by the haze value (H), which is the ratio of the diffuse transmittance (transmittance only including diffuse transmission T_d) to the total transmittance (i.e., transmittance including both specular and diffuse transmissions T_t) (Fig. 5-11).

$$H = \frac{T_d}{T_t} \times 100$$

The haze value of a perfectly transparent material is zero and increases as the degree of cloudiness increases.

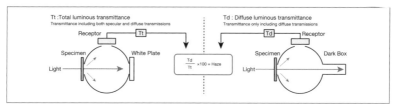

Fig. 5-11 Schematic diagram of haze measurement.

When measuring the total transmission, the efficiency (the ratio of the areas of apertures to the internal area of the sphere) of the integrating sphere (an apparatus which makes uniform incident light by diffusion) becomes important. The efficiency depends on the internal area of the sphere, the number of apertures and the method of covering the apertures. Differences in the size of the integrating sphere and the apertures will change the efficiency of the integrating sphere and cause errors in the measurement. However, by using a compensating aperture and conducting measurements, as shown in Fig. 5-12, it is possible to calibrate the changes in the efficiency of the integrating sphere. Therefore, there will be no difference in the measured value between devices. The current JIS standard (JIS K 7136) is based on this method.

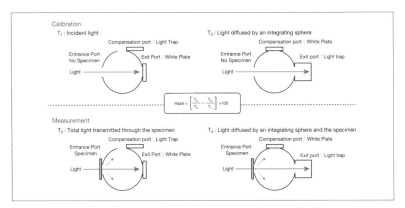

Fig. 5-12 Schematic diagram of haze measurement based on the current JIS standard.

4　Film thickness

There are various ways of measuring the thickness of a photocatalytic film, depending on its thickness, the shape and size of the sample, and the accuracy of the measurement.

If the film is sufficiently thick, the thickness can be calculated from weight measurements. From the mass of the film (w [g]) obtained by subtracting the mass of a base material alone, the area of a base material (A [cm^2]) and the density of the photocatalytic film (ρ [g cm^{-3}]), assuming that the photocatalytic film is coated evenly, the thickness (t [cm^{-3}]) can be calculated by the following equation.

$$t = \frac{w}{A\rho}$$

The density of the photocatalyst membrane is based on the literature value of the crystal. In the case of porous films, because the actual density is different from the literature value, estimated film thickness has an error. However, if this can be ignored, the method can be used for all samples for which weight can be measured.

Direct measurement with a micrometer is also possible. After measuring the thickness of the entire sample, the thickness of the photocatalyst layer can be determined by subtracting the thickness of the substrate (base material). However, these methods become difficult when the film thickness is thin, e.g., on the order of nanometers.

To measure the thickness of a thin film, if the sample (a base material) is a smooth substrate, the difference between the uncoated surface and the surface of the film can be measured. If the entire surface of the substrate is covered by the film and there is no step, the step can be created by mechanical scraping. Step measurements can be made using a probe-type step profiler (stylus profiler) or an atomic force microscope (AFM). In both methods, the up-and-down motion of the probe (tip) is recorded and the height of the step is measured. In principle, the AFM can measure film thicknesses of less than 1 nm; however, the smoothness of the substrate is also an issue in order to achieve such high accuracy.

Another method of measuring the thickness of a film is to taking a cross-sectional scanning electron microscope (SEM) image of a sample piece. Because an electron beam is irradiated to the sample during SEM measurement, if the sample is an insulator, it gradually becomes negatively charged (this is known as charge-up), making it difficult to photograph. In order to avoid this, it is necessary to make the sample conductive by sputtering (metal deposition). Although the magnification is lower, low vacuum SEM (LV-SEM) can be used to examine insulating samples, because electrostatic charging becomes less severe.

Interferometry is another method of measuring film thickness. When a thin film of titanium dioxide with a uniform thickness is prepared, blue, green or red coloring can be seen. This is called interference coloring and is caused by the repeated reflection of light with wavelength that is similar to the film thickness at the interface between the two sides of the film, where the refractive index varies. When the transmission or reflection spectrum of a thin film is measured, interference fringes appear, as shown in Fig. 5-13. The thickness of the thin film (d [nm]) can be calculated using the wavelengths λ_1 and λ_2 (nm) of the neighboring peaks (or troughs) and their refractive index n, as follows.

$$d = \frac{1}{2n} \left(\frac{1}{\lambda_2} - \frac{1}{\lambda_1} \right)^{-1} = \frac{\lambda_1 \lambda_2}{2n(\lambda_1 - \lambda_2)}$$

Here, the literal value of the crystal is used as the refractive index n. In the case of titanium dioxide, the refractive indices of rutile, anatase and brookite are 2.72, 2.52 and 2.63 respectively. For example, in an anatase titanium dioxide thin film with a refractive index of 2.52, if the peaks (or troughs) appear at 400 and 800 nm, the thickness of the film is about 160 nm. Because the distance between the stripes becomes longer with decreasing film thickness, and around 800 – 900 nm is the upper limit of the wavelength in common UV-visible spectrometers, if the film thickness of an anatase titanium dioxide thin film is less than 160 nm, two peaks (or troughs) cannot be observed. In this case, a spectrometer that can measure a much longer wavelength region is required.

Other methods of measuring film thickness include spectroscopic ellipsometry, where the change in polarization of the incident and reflected light is measured at each wavelength and the thickness is calculated from the measured data, and step measurements using confocal laser microscopy.

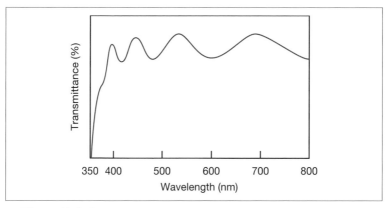

Fig. 5-13 Schematic of transmission spectrum of titanium dioxide thin film.

5 Pencil Hardness

The pencil hardness test is a method of determining which hardness of pencil will damage a coating, when a pencil of known hardness is pressed against it and scratched under certain conditions. The test method is standardized in JIS K5600-5-4 (ISO/DIN 15184).

The lead of a pencil consists of clay and graphite and becomes softer as the amount of graphite increases. The pencils used in the test have 14 hardness levels from 6B to 6H (6B, 5B, 4B, 3B, 2B, B, HB, F, H, 2H, 3H, 4H, 5H and 6H), with 6B being the softest and 6H being the hardest. The pencils tested by the Japan Paint Inspection Association (JPA) should be used, because the hardness of the lead varies slightly among manufacturers and production lots.

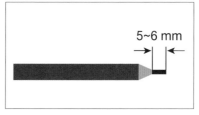

Fig. 5-14 Shape of the pencil tip.

The pencils are not sharpened at the tip, but are made into cylinders by removing the surrounding wood in order not to damage the lead, and flattening the tip with abrasive paper, as shown in Fig. 5-14.

Using a test apparatus as shown in Fig. 5-15, the pencil is placed at 45 ± 1° to the test surface, and the test surface is scratched to a length of at least 7 mm with a load of 750 ± 10 g and a speed of 0.5 to 1 mm/s. The presence of scratches is examined and the pencil hardness of the coating is determined from the hardness of the hardest pencil that does not leave a scratch.

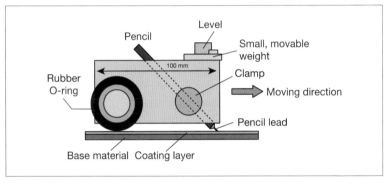

Fig. 5-15 Schematic diagram of pencil hardness tester.

References

1. Tsukaho Yahagi (Editor and Author), Hisashi Setoyama (Author), "Practical Analysis of Analytical Chemistry and Instrumental Analysis: Basics, Pretreatment, Method Selection, and Record Keeping" (Nikkan Kogyo Shimbun, 2020)
2. Kanagawa Institute of Industrial Science and Technology HP

(Norihiro Suzuki, Kengo Hamada)

Chapter **6**

Evaluation of Photocatalytic Performance

Importance of performance evaluation

Nowadays, various products using photocatalysis are being produced, for example, air purifiers. However, no matter how much you advertise that "it cleans the air," people will not believe it unless you show them the data that malodorous chemicals and viruses are actually removed. In other words, for research and development of photocatalytic products, how to evaluate the performance is a major point. Those who are familiar with photocatalysis may think "The JIS test has already been established as a performance evaluation method." However, environmental risks are becoming more and more diversified and serious, and evaluation methods need to evolve accordingly. In this Section, we introduce the outline of the conventional JIS test and the new performance evaluation method developed on the basis of it.

Outline of JIS test

JIS stands for "Japanese Industrial Standards," which are the standard accepted in Japan. The evaluation method of air purification performance (deodorization/odorizing performance) of photocatalytic materials is standardized by JIS/ISO. Table 1 shows the JIS and ISO test methods for photocatalytic materials. Performance to clean air and water, performance versus bacteria, molds and viruses, self-cleaning performance, etc. are established and updated in detail according to the application and decomposition target.

As an example, the air purification performance tests (R1701-1 to 5 and R1751-1 to 5) are described below. Both of these tests cover malodorous components and toxic substances such as toluene that are problematic in the actual environment (a description of each component is given below in Table 6-1). These substances are oxidized and decomposed by photocatalytic reactions, and carbon (C) in the molecules is converted to carbon dioxide (CO_2), hydrogen (H) is converted to (H_2O), and nitrogen (N) is converted to nitrate ion (NO_3^-). Therefore, in JIS and ISO tests, the performance of a photocatalysis is quantified by measuring how much each substance supplied at a certain concentration is decreased by the photocatalytic reaction and how much carbon dioxide, etc., which are decomposition products, are generated instead.

Table 6-1 JIS/ISO establishment status of photocatalysis performance evaluation test methods.

(Year of establishment or latest revision in parentheses)

Categorization	test laboratory (Appendix)	Test method	UV responsive type JIS No.	UV responsive type ISO No.	Visible light responsive type JIS No.	Visible light responsive type ISO No.
Air purification (Distribution)	A, D, E, F	NO$_x$	R 1701-1 (2016)	ISO 22197-1 (2016)	R 1751-1 (2013)	ISO 17168-1 (2018)
	A, E, F	Acetaldehyde	R 1701-2 (2016)	ISO 22197-2 (2019)	R 1751-2 (2013)	ISO 17168-2 (2018)
	A, E	Toluene	R 1701-3 (2016)	ISO 22197-3 (2019)	R 1751-3 (2013)	ISO 17168-3 (2018)
	A, E, F	Formaldehyde	R 1701-4 (2016)	ISO 22197-4 (2013) DIS 22197-4	R 1751-4 (2013)	ISO 17168-4 (2018)
	B	Methyl mercaptan	R 1701-5 (2016)	ISO 22197-5 (2013) DIS 22197-5	R 1751-5 (2013)	ISO 17168-5 (2018)
Air purification (Chamber)	-	Formaldehyde	-	-	R 1751-6 (2020)	ISO 18560-1 (2014)
anti-microbial	C, G, H, I, J, K	Antimicrobial	R 1702 (2020)	ISO 27447 (2019)	R 1752 (2020)	ISO 17094 (2014)
	-	Real environment antimicrobial (semi-dry method)	-	-	-	ISO 22551 (2020)
	-	Anti-mold	R 1705 (2016)	ISO 13125 (2013)	-	-
	-	Anti-algae	-	ISO 19635 (2016)	-	-
	C, G, J, K, L	Anti-virus	R 1706 (2020)	ISO 18061 (2014)	R 1756 (2020)	ISO 18071 (2016)
Self-Cleaning	A	Water contact angle	R 1703-1 (2020)	ISO 27448 (2009)	R 1753 (2013)	ISO 19810 (2017)
	A	Methylene blue	R 1703-2 (2014)	ISO 10678 (2010)	-	-
	-	Resazurin ink	-	ISO 21066 (2018)	-	-
Water quality	B	Dimethyl sulfoxide	R 1704 (2007)	ISO 10676 (2010)	-	-
Complete decomposition	B	Acetaldehyde	-	-	R 1757 (2020)	ISO 19652 (2018)
Oxidation reaction activity (in water method)	-	Dissolved oxygen (phenol decomposition)	R 1708 (2016)	ISO 19722 (2017)	-	-
	-	Total organic carbon (TOC)	R 1711 (2019)	ISO 22601 (2019)	-	-
light source	A	Standard light source	R 1709 (2014)	ISO 10677 (2011)	R 1750 (2012)	ISO 14605 (2013)
	-	Standard LED light source	-	-	-	DIS 24448

- **Acetaldehyde:** chemical formula CH_3CHO, produced by the metabolism of alcohol, is generally regarded as the cause of hangovers, and is also said to increase dependence on tobacco, and is carcinogenic.
- **Toluene:** Chemical formula $C_6H_5CH_3$, a kind of organic solvent, used for paints and adhesives. Toluene is used as a solvent in building materials and is sometimes released indoors, and is considered to be one of the substances causing the "sick building" syndrome. Inhalation of the vapor is harmful to human health, and repeated inhalation over a long period of time can cause irreversible brain damage.
- **Formaldehyde:** Chemical formula $HCHO$, when dissolved in water, it becomes formalin. Formaldehyde is widely used as a raw material for various resins, and is considered to be one of the causative agents of the "sick building" syndrome as well as toluene. Formaldehyde is widely used as a raw material for building materials, etc., and is slowly released from these materials.
- **Methyl mercaptan:** Chemical formula CH_3SH, is a colorless gas that smells like rotten onions and is one of the causes of bad breath.

In the air purification performance tests (R1701-1 to 5 and R1751-1 to 5), the reactors used are all the same, but the gas species and concentrations to be decomposed and the wavelength and intensity of the light irradiated are different. Figure 6-1 shows an overview of the JIS test (R1701-2) of UV-responsive photocatalysts using acetaldehyde. A sample of a given size is placed in a reactor, and the decomposition behavior of acetaldehyde is evaluated at a given concentration, flow rate, temperature, humidity and light intensity. In other words, the JIS test can be performed if the following equipment and instruments are available.

① Gas flow part:
Standard gas cylinder, pure air cylinder, gas mixing device, humidity controller, tubing and valves;

② Photocatalytic reaction:
Reactor, light source (black light/fluorescent lamp + UV-cut filter,

light source (black light/fluorescent lamp + UV-cut filter, see Chapter
7), light meter, thermostatic bath, tubing and valves;

③ Analysis of unreacted gas and reaction products:

Gas chromatograph*, ion chromatograph (for nitrogen oxide
decomposition test).

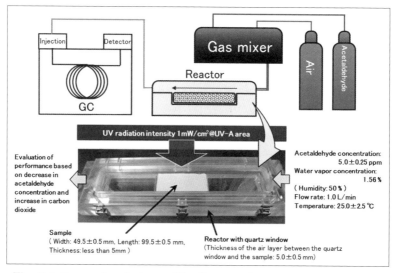

Fig. 6-1 Example of air purification performance test method (JIS R 1701-2 Acetaldehyde removal test).

＊Gas chromatograph: A device for separating and examining the components of a gas or liquid to determine how much of each component is contained. A column is a narrow tube in which a specific component is adsorbed. When the mixture is injected into the column, the speed at which each component passes through the column changes, so that the components are gradually separated, as shown in the diagram below. The separated components are then analyzed and quantified by the detector. A device for analyzing gases is called a gas chromatograph (GC), a device for analyzing liquids is called a high-performance liquid chromatograph (HPLC), and a device for analyzing ions in liquids is called an ion chromatograph (IC).

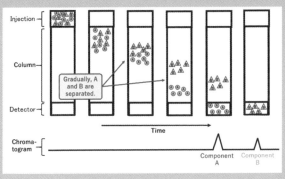

Figure 6-2 shows an example of the JIS R 1701-2 acetaldehyde removal performance test results. The gas at the outlet of the reactor was sampled, and the concentrations of acetaldehyde and carbon dioxide were analyzed by use of the gas chromatograph. When the UV irradiation is started 40 minutes after the start of the test, 5 ppm* of acetaldehyde poured into the reactor is decomposed on the surface of the photocatalyst sample and the concentration of acetaldehyde at the reactor outlet decreases. After 3 hours, when the UV irradiation is stopped, the acetaldehyde concentration returns to the supplied concentration of 5 ppm. The greater the decrease in acetaldehyde and the increase in carbon dioxide, the better the photocatalyst's air purification performance. When 1 ppm of acetaldehyde is completely decomposed, 2 ppm of carbon dioxide is produced. In the case of Fig. 6-2, the decrease of acetaldehyde is about 3.5 ppm and the increase of carbon dioxide is about 7 ppm, which means that almost 100% of the decreased acetaldehyde is completely decomposed to carbon dioxide.

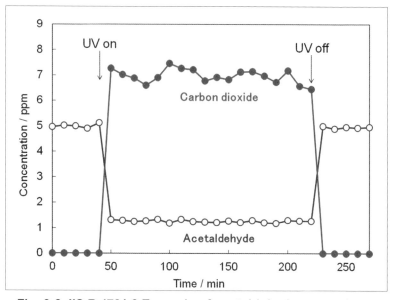

Fig. 6-2 JIS R 1701-2 Example of acetaldehyde removal test results.

＊1 ppm = 0.0001% is used to express a very small percentage.

Other tests can also be carried out by combining the necessary equipment and instruments, referring to the test methods specified in the JIS standards (available on the JSA GROUP Webdesk, the web sales site of the Japanese Standards Association) respectively. At the end of the chapter we have compiled a list of institutions that can carry out JIS tests and their contact details. As shown in Fig. 6-3, KISTEC provides total support for photocatalysis and is a recommended testing institute by the Photocatalysis Industry Association of Japan. KISTEC provides total support for photocatalysis, as shown in Fig. 6-3, and is also a recommended testing institute by the Photocatalysis Industry Association of Japan (PIAJ). PIAJ-certified products are listed on the website of the Photocatalysis Industry Association of Japan (see Chapter 14) with detailed JIS test data, so that companies developing new products can compare the level of their products according to their JIS test results.

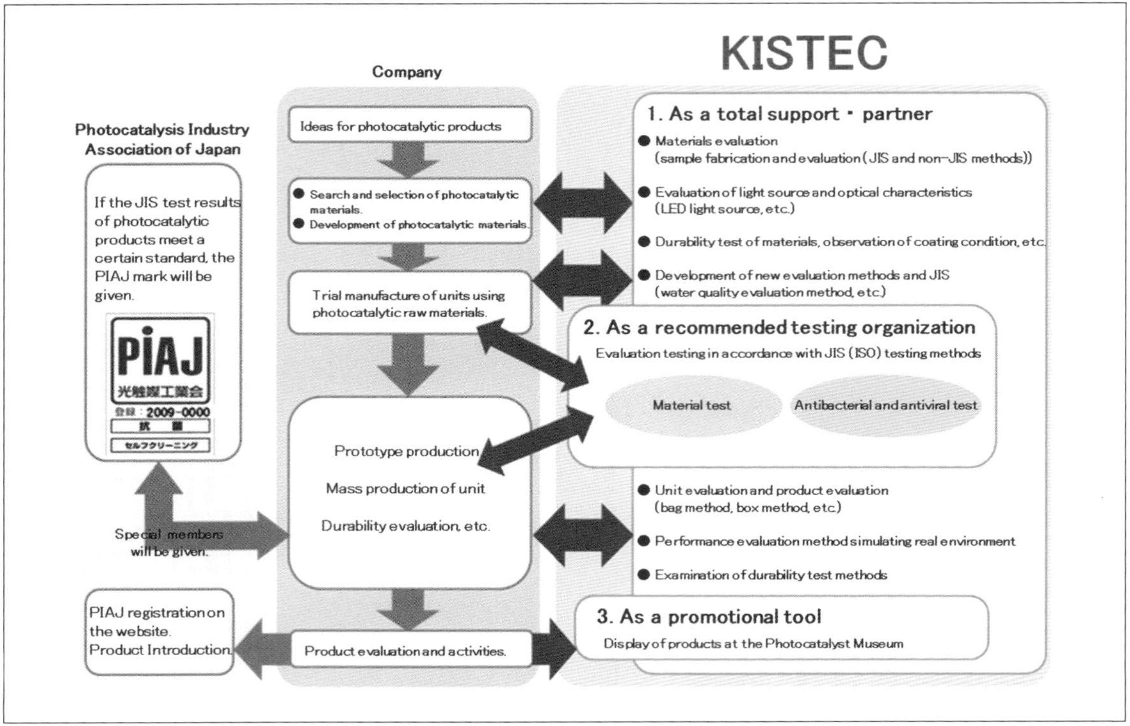

KISTEC

1. As a total support・partner

- Materials evaluation
  (sample fabrication and evaluation (JIS and non-JIS methods))
- Evaluation of light source and optical characteristics
  (LED light source, etc.)
- Durability test of materials, observation of coating condition, etc.
- Development of new evaluation methods and JIS
  (water quality evaluation method, etc.)

2. As a recommended testing organization

Evaluation testing in accordance with JIS (ISO) testing methods

Material test   Antibacterial and antiviral test

- Unit evaluation and product evaluation
  (bag method, box method, etc.)
- Performance evaluation method simulating real environment
- Examination of durability test methods

3. As a promotional tool

Display of products at the Photocatalyst Museum

Company

Ideas for photocatalytic products

- Search and selection of photocatalytic materials.
- Development of photocatalytic materials.

Trial manufacture of units using photocatalytic raw materials.

Prototype production

Mass production of unit

Durability evaluation, etc.

Product evaluation and activities.

Photocatalysis Industry Association of Japan

If the JIS test results of photocatalytic products meet a certain standard, the PIAJ mark will be given.

PiAJ
光触媒工業会
登録：2009-0000
氏　名
セルフクリーニング

Special members will be given.

PIAJ registration on the website.
Product Introduction.

Fig. 6-3 Image of "Photocatalysis total support" implemented by KISTEC.

Evaluation method of decomposition performance for various VOC by applying JIS standard

With the diversification of air pollution problems in recent years, there is a growing need to test under conditions closer to the actual environment, in addition to the JIS test described in the previous Section. In this Section, we introduce the results of decomposition tests for volatile organic compounds (VOCs), which are not covered by the JIS test but are regarded as a significant problem in the environment. Using the JIS test reactor shown in Fig. 6-1, we tested the decomposition of various VOCs (methyl ethyl ketone, n-hexane, ethyl acetate, and 2-propanol, all of which are commonly used as organic solvents. Fig. 6-4 shows the results of decomposition of the various VOCs ; the simplified chemical structures are shown in Fig. 6-4) by the photocatalytic sample after adjusting the concentration to 5 ppm, the same as that of acetaldehyde, and introducing them into the reactor. Figure 6-4 shows the results of the photocatalytic decomposition of acetaldehyde and carbon dioxide at the outlet of the reactor.

It can be seen that the decomposition behavior differs significantly depending on the type of VOC. The reason for this is the difference in the adsorption properties on the photocatalyst surface. The photocatalytic decomposition of organic substances takes place almost exclusively on the surface, with adsorption on the photocatalyst surface

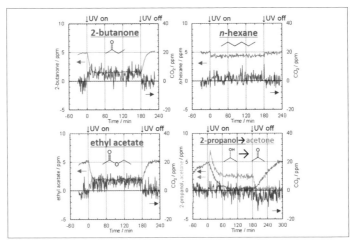

Fig. 6-4 Results of decomposition tests of various VOCs (methyl ethyl ketone, n-hexane, ethyl acetate, and 2-propanol) using JIS test reactors.

first, followed by oxidation by holes and reactive oxygen species (see adsorption mechanism, Chapter 5). Therefore, the removal rate of methyl ethyl ketone and ethyl acetate, which are relatively easy to adsorb, is high, but the removal rate of n-hexane, which is difficult to adsorb, is low. Some compounds, such as 2-propanol, have a high removal rate but do not decompose to the final decomposition product, carbon dioxide. This may also be due to the adsorption of the decomposition products on the photocatalyst surface. In other words, although photocatalysis is often explained so that it appears that it can decompose any type of organic matter, there are some compounds that are difficult to decompose or do not decompose completely, depending on experimental conditions and the environment.

Demonstration photocatalytic filter performance tester

The JIS test uses a relatively small sample of (5 × 10 cm) and evaluates the performance in special reactors under specific conditions. Therefore, the situation differs greatly from that of actual air purifiers, and even if a good result is obtained in the JIS test, the expected performance may not be obtained in the final product. Therefore, we constructed a prototype of the demonstration photocatalytic filter test machine (Fig. 6-5), which can be used for various sizes and conditions, and evaluated the performance of the A4 size photocatalytic filter. This demonstration photocatalytic filter tester can change the effective area of the filter, the

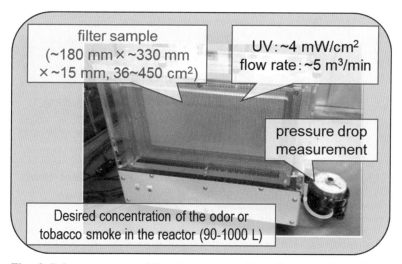

filter sample
(~180 mm × ~330 mm
× ~15 mm, 36~450 cm²)

**UV: ~4 mW/cm²
flow rate: ~5 m³/min**

pressure drop measurement

Desired concentration of the odor or tobacco smoke in the reactor (90-1000 L)

Fig. 6-5 Appearance of the demonstration photocatalytic filter testing machine and settable conditions.

amount of light, and the amount of airflow, so it can create a situation similar to the design conditions of an actual air purifier.

Fig. 6-6 shows the results of decomposition tests of 50 ppm acetaldehyde gas in 90-L and 1000-L reactors, respectively (UV intensity 4.0 mW/cm², airflow 5.4 m³/min, filter effective area 150 × 240 mm). If you look closely at the curves of the concentration change, you will see that the time taken for the concentration to decrease by half (half-life) is constant. Such a curve can be analyzed using the mathematical concept of an exponential function. Figure 6-7 shows an image of this exponential function.

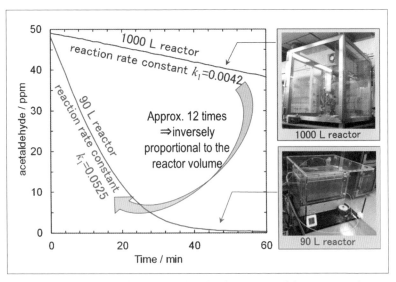

Fig. 6-6 Results of acetaldehyde decomposition test using demonstration-type photocatalytic filter tester.

Using this exponential function, we can analyze the results of Fig. 6-7 and calculate the rate constants for each reaction, thus quantifying the speed of the reaction. As a result, it was observed that the decomposition was 12 times faster in the 90 L reactor than in the 1000 L reactor. This is inversely proportional to the volume ratio of the reactor, which is a reasonable result.

In the acetaldehyde decomposition performance test using the demonstration photocatalytic filter tester, it was confirmed that the decomposition behavior of the malodorous components changed in response to changes in the test conditions. In other words, it is possible to evaluate the performance according to various filter sizes and usage conditions, and it can be said that the design provides data that can be used as a reference in making design policies for actual air purifiers.

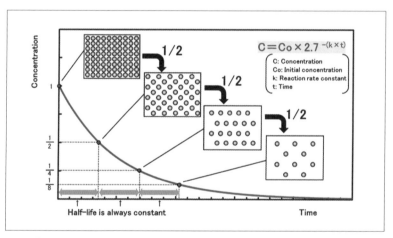

Fig. 6-7 Schematic diagram of exponential functions.

Conclusion

In this Chapter, we introduced the outline of the JIS/ISO test and applied evaluation method based on it. Related manufacturers and public testing laboratories are actively engaged in basic study and standardization projects related to photocatalytic materials and processed products, and are working on the revision of existing JIS and JIS/ISO methods and the establishment of new test methods. In today's world, where environmental risks are becoming increasingly diverse and serious, it is important to match the needs and seeds of industry and to promote them to society. KISTEC, in collaboration with Tokyo Institute of Technology and Nara Medical University, has confirmed that visible light-responsive photocatalysts can inactivate the novel coronavirus (announced on 25 September 2020). Furthermore, we have made preparations to use this know-how to undertake actual tests using coronaviruses. In the future, it will be necessary to have a system in which the fields of industry, academia, the public sector and medicine work together to support the further development of photocatalytic applications through research and development of appropriate evaluation methods.

(Tsuyoshi Ochiai)

Institutions that can perform JIS testing and their contact information

Sign	Institution name	Address · Department	Phone number
A	Kanagawa Institute of Industrial Science and Technology. Kawasaki Technical Support Department	KSP East 1 F, 3-2-1 Sakado, Takatsu-ku, Kawasaki-shi, Kanagawa Photovoltaic Evaluation Group	044-819-2105 FAX: 044-819-2108
B	Kanagawa Institute of Industrial Science and Technology. Kawasaki Technical Support Department	KSP East 1 F, 3-2-1 Sakado, Takatsu-ku, Kawasaki-shi, Kanagawa Materials Analysis Group	044-819-2105 FAX: 044-819-2108
C	Kanagawa Institute of Industrial Science and Technology. Antimicrobial Testing Laboratory	LiSE 4 c-2, 3-25-13 Tonomachi, Kawasaki-ku, Kawasaki-shi, Kanagawa Photocatalysis Group, Antibacterial and Antiviral Research Group	044-280-1181 FAX: 044-280-1182
D	Environmental Technical Laboratory Ltd.	2-11-17, Kohoku, Adachi-ku, Tokyo Research & Analysis Dept.	03-3898-6643 FAX: 03-3890-3086
E	Environmental Management & Technology Center	2-9-10 Kawaguchi, Nishi-ku, Osaka City Survey Section, Environmental Engineering Department	06-6583-7122 FAX: 06-6583-3274
F	Chemicals Evaluation and Research Institute, Japan	1600 Shimotakano, Sugito-machi, Kitakatsushika-gun, Saitama Chemical Standard Department Technical Section 1	0480-37-2601 FAX: 0480-37-2521
G	Japan Food Research Laboratories	Saito Research Institute, 7-4-41, Saito Asagi, Ibaraki, Osaka, Japan Microorganism Research Section, Microorganism Department	072-641-8954 FAX: 072-641-8965
H	KAKEN Test Center	2-5-19 Edobori, Nishi-ku, Osaka City, Osaka Biology Laboratory	06-6441-0399
I	BOKEN QUALITY EVALUATION INSTITUTE	1-6-24 Chikko, Minato-ku, Osaka City, Osaka Osaka Functional Testing Center	06-6577-0157
J	Analytical Technology Center, General Research Laboratory, TOTO Ltd.	2-8-1 Honmura, Chigasaki-shi, Kanagawa Analytical Technology Center, General Research Laboratory	0467-54-3595 FAX: 0467-54-1185
K	Kitasato Research Center for Environmental Science	1-15-1 Kitasato, Minami-ku, Sagamihara-shi, Kanagawa Microorganism Department Biotech Section	042-778-8324
L	Japan Textile Products Quality and Technology Center (QTEC)	5-7-3 Shimoyamate-dori, Chuo-ku, Kobe Kobe Testing Center	078-351-1891 FAX: 078-351-1894

(Based on the list of recommended testing institutions on the Photocatalysis Industry Association of Japan website. https://www.piaj.gr.jp/registered_products/institution/)

地方独立行政法人
KISTEC 神奈川県立産業技術総合研究所

Chapter **7**

Light Source Systems
(Wavelength Characteristics, Intensity, Lifetime, Price, etc.)

Sunlight
Tungsten lamps
Mercury lamps
Xenon lamps
Light-emitting diodes (LEDs)
Lasers
Artificial solar lamps

Chapter 7

Light sources

1 Sunlight

Although sunlight includes X-rays with very short wavelengths and radio waves with wavelengths of several hundred meters, most of sunlight is in the ultraviolet, visible and infrared range. As seen from the spectrum of sunlight shown in Fig. 7-1, about half of the spectrum is in the visible light range (wavelength 400-800 nm) and most of the rest is in the infrared (wavelength > 800 nm) range. Ultraviolet rays with 300 nm or shorter wavelength are nearly nonexistent, and ultraviolet rays with approximately 400 nm or shorter wavelength, which can be absorbed by titanium dioxide, account for about only 3-4% of the total. There is a drop in the intensity of light at certain wavelengths, because the stratospheric ozone layer absorbs this light. Therefore, the amount of light that reaches the Earth is decreased.

Although the intensity of light emitted from the Sun is nearly constant throughout the year, the sunlight intensity on the Earth's surface changes with the season due to astronomical factors such as the Earth's elliptical orbit and the tilt of the Earth's axis. In addition, because 30% of the sunlight returns into space by reflecting from the Earth's surface and clouds, and scattering in the atmosphere, only 70% reaches the Earth, and its light intensity is about 1 kW/m^2.

Although sunlight is attractive as a free and inexhaustible source of light (the sun has another 5 billion years to live), needless to say, it is not available in bad weather or at night. Another problem is that both the light intensity and wavelength distribution are highly variable. It also contains a significant amount of infrared radiation, which means that measures must be taken to prevent the heating of the reaction system.

A particular problem of sunlight is the low energy density. Therefore, it is necessary to consider the use of reflectors and lenses to concentrate the light, or devices that track the movement of the sun. However, this is not a major problem as long as a sufficiently large light-receiving area is available and large equipment is acceptable.

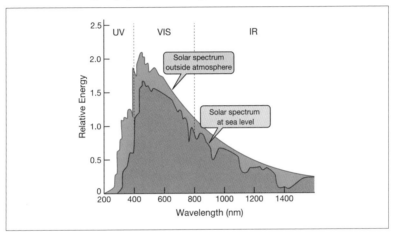

Fig. 7-1 Spectrum of sunlight.

2 Tungsten lamps

A tungsten lamp is an incandescent lamp with a tungsten filament. The temperature of the filament rises to around $2,500°$ C to $2,650°$ C and the light is emitted by blackbody radiation. The light is continuous, covering the visible to near-infrared range, and has a reddish color.

The filament evaporates due to its own heat, causing the filament to lose weight and break, or the evaporated tungsten adheres to the tube wall and blackens the tube wall, resulting in a reduction in light output. Therefore, in order to prevent tungsten from evaporating, an inert gas such as argon or nitrogen is usually filled in the tube.

In addition to the inert gas, some lamps contain small amounts of halogen elements (F, Cl, Br, I) and are called "halogen lamps." In a halogen lamp, the tungsten evaporated from the filament combines with the halogen atoms in the enclosure to form tungsten halide. This molecule separates into halogen atoms and tungsten molecules near the hot filament. The tungsten atom returns to the filament, while the free halogen atom repeats the previous reaction (Fig. 7-2). This series of reactions is called the "halogen cycle," which inhibits the blackening of the tube wall and prevents the filament from wearing out.

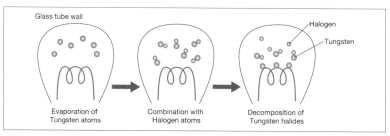

Fig. 7-2 Schematic diagram of the halogen cycle.

3 Mercury lamps

A mercury lamp is a light source that uses the light radiation generated by the discharge of mercury vapor in a glass tube. Mercury lamps are classified into low-pressure and high-pressure lamps according to the degree of mercury vapor pressure inside.

Low-pressure mercury lamps have an internal mercury vapor pressure of about 0.8 Pa (about 0.8×10^{-6} atm) when lit and use a glow discharge (continuous discharge). They are line-spectrum light sources that emit intense ultraviolet light at wavelengths of 184.9 nm and 253.7 nm, which are known as the resonance lines of mercury (Fig. 7-3 (a)). Depending on the use, the light wavelength is adjusted by changing the glass material of the lamp or using fluorescent materials.

When high-purity synthetic quartz glass is used as the glass material, 184.9 nm wavelength light can be emitted through the glass. This lamp is called an "ozone lamp," because ozone is produced when the oxygen in the air absorbs light of this wavelength. Due to the use of special glass and the high cost, these lamps are rarely used unless chemical reaction of ozone as well as photocatalytic reaction is demanded. In contrast, if ordinary quartz glass is used as the glass material, 184.9 nm wavelength light is absorbed by the glass; thus, 253.7-nm wavelength light is mainly emitted. This lamp is called a "germicidal lamp" because this wavelength is known to have a high germicidal effect.

Another type of lamp that uses a fluorescent substance to adjust the wavelength is the "black light." A fluorescent substance with an emission maximum around 350 nm is coated on the inside of the glass, and the 253.7-nm wavelength light is converted to this longer wavelength light. Colored glass is also used to obtain only ultraviolet light with a wavelength of 300-400 nm by blocking out visible light. Lamps using ordinary glass instead of colored glass are

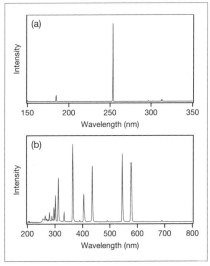

Fig. 7-3 Emission spectrum of (a) low-pressure and (b) high-pressure mercury lamps.

(provided by ORC Manufacturing Co., Ltd.)

also commercially available as photochemical fluorescent lamps (chemical lamps). The only difference between the two lamps is that a small amount of visible light is included in the chemical lamp, but the wavelength distribution and intensity in the UV region is nearly identical. Chemical lamps are much cheaper than black lights, because the cost of the lamp is determined by the glass material.

Fluorescent lamps for lighting purposes are commercially available coated with a variety of fluorescent materials, with the main difference being in the visible light range. It should be noted that, just because a photocatalytic reaction occurs under fluorescent lamp illumination, this does not prove that the reaction occurs under visible light, because fluorescent lamps also contain ultraviolet light below 400 nm wavelength, though it is a small amount.

High-pressure mercury lamps have a mercury vapor pressure of around 100 kPa to 1,000 kPa (1-10 atm) when lit and are lamps that use arc discharge (a continuous discharge that produces intense light and heat). Generally, "mercury lamp" refers to this high-pressure type. The emission spectrum is shown in Fig. 7-3(b), which shows a large number of emission lines in the ultraviolet and visible regions.

Most of the commercial lamps for lighting are air-cooled and can be purchased for a few thousand yen. In contrast, those for photochemical use have a quartz outer tube and often have a water-cooling system between the emission part and the outer tube; thus they cost several hundred thousand yen.

In order to light a mercury lamp, a voltage must be applied to keep the electrodes warm until the mercury vapor pressure reaches a sufficiently high level; therefore, the lamp does not light immediately after being switched on. Several minutes are required until the light intensity stabilizes. In addition, because the illuminated lamp has high mercury vapor pressure and temperature inside, the discharge for relighting it is impossible immediately after turning it off. To relight it, both the mercury vapor pressure and temperature should decrease with time.

The manufacture, import and export of mercury lamps will be banned from 2021 onwards under the Minamata Convention on Mercury, except for specially permitted cases.

4 Xenon lamps

Xenon lamps are discharge lamps that emit light through the excitation of xenon gas by an arc discharge (a contentious discharge that produces intense light and heat). Xenon lamps are characterized by their high luminosity and the stable emission spectrum, regardless of a fluctuation of power input and lifetime. Unlike mercury lamps, because there is no need for vapor production, the output is instantaneous and stable.

They are used as light sources in solar simulators, because the wavelength distribution of the emission spectrum in the UV to visible range resembles that of sunlight. The light intensity is nearly constant in the wavelength range of 400 to 800 nm, and the shape of the emitting part is similar to that of a point (point light source), making it ideal as a light source for spectrographs in which the light is split into different wavelengths.

The basic cooling system uses air and a lamp housing. Some lamps have a reflector built into the lamp itself in order to focus the light efficiently.

5 | Light-emitting diodes (LEDs)

A light-emitting diode is a semiconductor device that emits light by passing a current through a p/n junction. The basic mechanism is summarized in Fig. 7-4. When a "forward" voltage is applied, holes and electrons move towards the p/n junction, where they combine and disappear. The electrons then transit from a high energy state to a low energy state, and the excess energy is emitted as light.

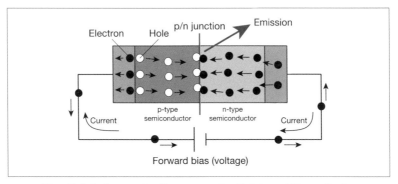

Fig. 7-4 Schematic diagram of a light-emitting diode.

The combination of holes and electrons at the p/n junction occurs when the electrons fall from the conduction band, with higher energy, to the valence band, with lower energy. The energy difference (band gap) is therefore roughly equivalent to the energy released as light. This is why LEDs are characterized by their high monochromatic properties (narrow emission spectrum of 10-20 nm).

The wider the band gap, the higher the energy of the light emitted (i.e., light with shorter wavelength). Because the value of the band gap differs among semiconductor materials, desired light-emitting diodes (LEDs) are made from materials with a band gap that matches the color of light to be emitted (white light can be obtained by combining blue or UV LEDs with fluorescent materials, or by combining red, green and blue LEDs).

Unlike conventional mercury lamps, UV LEDs emit light and can be switched on and off immediately; therefore, they do not need to keep being relit. They are an environmentally friendly light source, because they do not contain harmful substances such as mercury and do not produce ozone. In addition, they are long-lasting light sources with low power consumption. Furthermore, they are smaller than existing light sources and emit less heat, allowing for a variety of flexible application designs. As a result, they are attracting attention as a replacement for existing light sources. Table 7-1 summarizes the manufacturers of UV LEDs (as of January 2021), including those from Korea and Taiwan. Many UV-A (315 nm to 400 nm in wavelength) LEDs are manufactured, partly because visible LED manufacturing equipment can be used, while UV-C (100 nm to 280 nm in wavelength) LEDs, which have shorter wavelengths and are more effective at sterilization, have come on the market.

Since photocatalysts require UV and visible light, it will be very important to evaluate the effectiveness of photocatalytic reaction achievable under LEDs as a light source. It is also important to select the proper LED for the performance evaluation. For this reason, work is currently underway to make visible LEDs as the standard light source used in the performance evaluation of photocatalytic materials, and we have already reached the halfway point of standardization as a draft international standard. The use of visible LEDs as a standard light source will become possible in the near future. In addition, the standardization of ultraviolet LEDs has also started, and this will be available as a standard light source in a few years.

Table 7-1 List of UV LED Manufacturers (as of January 2021).

Company name	Products URL	QR code
Takatsuki Electric Industry Co., Ltd.	https://www.takatsuki-denki.co.jp/products/index.html	
KYOTO SEMICONDUCTOR Co., Ltd.	https://www.kyosemi.co.jp/products/ked373us1/	
Uni-Technology.co.,ltd	http://www.uni-technology.co.jp/seihinjoho2.html	
KY TRADE CO., LTD.	https://ky-trade.co.jp/product/product_syodokei/syodokei_75/	
OPTO SCIENCE, INC.	https://www.optoscience.com/maker/crystal_is/	
Aitec System Co.,Ltd.	https://aitecsystem.co.jp/category/uv-led/dv-uv-led/	
NIKKISO CO., LTD.	https://www.nikkiso.co.jp/products/duv-led/products.html#ac06	
NICHIA CORPORATION	http://www.nichia.co.jp/jp/product/uvled.html	
Raytron CO., Ltd.	https://www.raytron-japan.co.jp/uv-led/	
PHOENIX Electric Co.,Ltd.	https://www.phoenix-elec.co.jp/product/led/	
Nitride Semiconductors Co.,Ltd.	http://www.nitride.co.jp/products/lineup.html	

Company name	Products URL	QR code
STANLEY ELECTRIC CO., LTD.	https://www.stanley-components.com/jp/product/ultraviolet.html	
KODENSHI CORP.	http://www.kodenshi.co.jp/news/2008/11/uv-led.html	
EL-Seed Corp.	http://elseed.com/jproducts/jthe-kamiyama-led/	
Ushio Inc.	http://www.ushio-optosemi.com/jp/products/led/	
ALPHA-ONE ELECTRONICS LTD.	https://www.alpha-one-el.com/products_01.html	
SEOUL VIOSYS CO., LTD.	https://www.businesswire.com/news/home/20180813005211/ja/	
Sensor Electronic Technology (SET)	http://www.s-et.com/en/product/lamp/	
Refond	http://www.refond.co.jp/uvled.html	
OptoSupply	http://www.optosupply.com/product/list2.asp?id=216	
Lumex	https://www.lumex.com/led-thru-hole.html?specs=26619	
Ligitek Electronics	https://www.ligitek.com	

Table 7-1 List of UV LED Manufacturers (as of January 2021)

Company name	Products URL	QR code
LEDtronics	https://www.ledtronics.com/products.aspx?page=18#undefined,-1	
Jenoptik	https://www.jenoptik.com/products/optoelectronic-systems/photodiodes-and-led/point-sources	
DOWA Electronics Materials Co., Ltd.	https://www.ultraviolet-led.com/wave/	
Crystal IS	https://www.klaran.com/products/uvc-leds	

6 | Lasers

Lasers are devices that produce light that is directional (travels straight with little or no spread), monochromatic (wavelength is uniform) and coherent (the peaks and troughs of the light waves are aligned).

Although there are many different types of lasers, here we will discuss semiconductor lasers, which have the same light-emitting mechanism as LEDs.

The basic structure of a semiconductor laser is shown in Fig. 7-5. The active layer (light-emitting layer) is sandwiched between n-type and p-type cladding layers (double heterostructure) on an n-type substrate, so that light is reflected from the edge of the active layer. When a forward voltage is applied, electrons from the n-type cladding layer and holes from the p-type cladding layer are injected into the active layer and recombine within the active layer to emit light. Although this light is not laser light yet, because the refractive index of the cladding layer is lower than that of the active layer, the light is confined within the active layer. As the two surfaces of the active layer act as reflective surfaces, the light is amplified as it travels back and forth through the active layer, causing induced emission (a phenomenon in which strong, phase-locked light is produced). In this way, laser light is emitted.

It is unlikely that a laser will be needed as a light source in many practical applications. On the other hand, their monochromatic nature and the high intensity of the light enables multiphoton absorption, which would be useful in basic research.

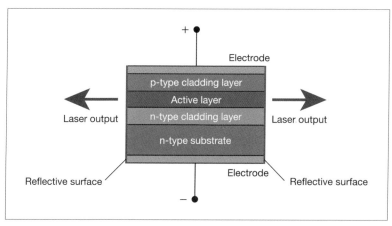

Fig. 7-5 Schematic diagram of a semiconductor laser.

7 | Artificial solar lamps

Light sources that produce light close to sunlight are also available on the market.

One example is shown below. SOLAX-NEXT [product name] (Fig. 7-6) uses LED lamps, and the wavelength distribution is flat in the visible light range, as shown in Fig. 7-7.

Other products are the SOLAX 100 W series and SOLAX 500 W series.

Fig. 7-6 Artificial solar lamp.

For further information, please contact

Ceric Corporation 6-3-8-703 Akasaka, Minato-ku, Tokyo TEL: 03-6807-4811

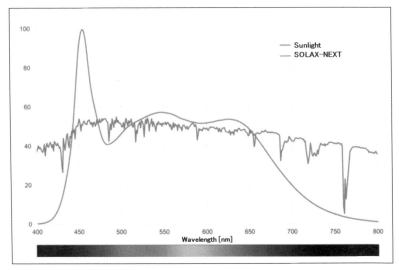

Fig. 7-7 Spectral distribution of sunlight and SOLAX-NEXT.

The spectral distributions of typical light sources are summarized in Fig. 7-8.

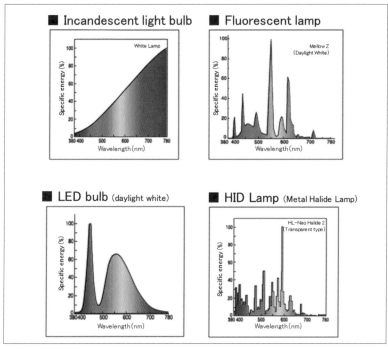

Fig. 7-8 Spectral distributions of light sources.
Source: TOSHIBA LIGHTING & TECHNOLOGY CORPORATION website

(Norihiro Suzuki)

Chapter 8

Equipment

In recent years, there have been increasing expectations for environmental purification using titanium dioxide photocatalysts. This is due to the increasing and diversifying environmental risks, such as air pollution and other environmental problems, the high incidence of infectious gastroenteritis caused by norovirus, and the spread of new types of influenza and new coronaviruses. However, the application of titanium dioxide photocatalysts to the field of environmental purification equipment is a technically challenging field, unlike the field of exterior materials, where the cleaning effect of superhydrophilicity is significant. In this section, we will explain the design guidelines for effective environmental purification systems, based on several examples.

1   Limitations of titanium dioxide photocatalysis, and design guidelines based thereon

The decomposition of substances as a result of titanium dioxide photocatalysis is triggered by the generation of excited electrons and holes by ultraviolet light, followed by the generation of reactive oxygen species. This is very different from conventional catalysis, where the reaction is thermally promoted at the active site. Therefore, although the photocatalytic reaction proceeds at relatively low temperatures, the reaction rate is significantly slower if sufficient UV light is not available. Figure 8-1 shows a simple estimation. The number of photons of ultraviolet radiation in sunlight falling on 1 cm^2 of the earth's surface is of the order of 10^{15} per second at most. Even if we assume that each photon is capable of decomposing one molecule on the surface (quantum yield 100%), it would take about 20 days to completely decompose a few drops of water (0.18 mL, 6.0×10^{21} molecules) on a 1 cm^2 titanium dioxide surface.

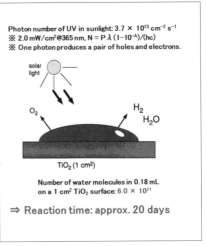

Photon number of UV in sunlight: 3.7×10^{15} cm^{-2} s^{-1}
※ 2.0 mW/cm^2@365 nm, $N = P \lambda (1-10^{-A})/(hc)$
※ One photon produces a pair of holes and electrons.

solar light

O_2

H_2
H_2O

TiO_2 (1 cm^2)

Number of water molecules in 0.18 mL
on a 1 cm^2 TiO$_2$ surface: 6.0×10^{21}

$\Rightarrow$ Reaction time: approx. 20 days

Fig. 8-1 Image of the reaction rate of a photocatalysis.

This means that photocatalytic reactions are in principle incapable of decomposing large quantities of pollutants in a short time. In particular, in the case of water purification, the pollutant blocks or scatters the light, so that not enough light reaches the surface of the photocatalyst, and the slow diffusion of the substance makes it difficult for the substance to be decomposed to reach the surface of the photocatalyst, which further reduces the reaction efficiency. In this section, we explain the following two points for creating an efficient environmental purification system using a photocatalysis, using specific examples, such as an air purifier.

(1) Design titanium dioxide photocatalyst support and reactor to maximize reaction area and mass transport efficiency.

(2) Combining titanium dioxide photocatalysis with other treatment technologies, such as electrolysis and ozonation, to take advantage of synergistic effects.

2 | Examples of effective design

Titanium dioxide photocatalyst generally exists in the form of fine particles with a diameter of several nm to several tens of nm, and, if it is used as it is to purify air or water, the titanium dioxide particles must be removed by a filter after treatment. Therefore, from the viewpoint of practicality, titanium dioxide particles are supported on the surface of a substrate to form a photocatalytic filter, which is used in combination with an ultraviolet light source to purify air and water (Fig. 8-2). For example, an air purifier consists of a pre-filter, a photocatalytic filter, an ultraviolet light source, and a fan. Harmful substances and microorganisms in the room are taken in by the fan, adsorbed on the surface of the photocatalytic filter, and oxidized and decomposed by the photocatalytic reaction caused by ultraviolet light irradiation. At present, photocatalytic filters in which titanium dioxide particles are heat-treated so as to adhere onto porous ceramic filters have been developed and put into practical use. Aiming for further application, manufacturers of home appliances and others are considering higher performance, smaller

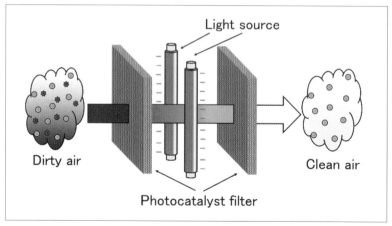

Fig. 8-2 Image of the reaction rate of a photocatalysis.

size, and cost reduction of photocatalytic filters and air purifiers incorporating them, and they are appearing on the market (see Chapter 9).

The Kanagawa Institute of Industrial Science and Technology (KISTEC, see Chapter 6), which evaluates and develops photocatalytic products, has developed a photocatalytic filter, TMiP (titanium mesh impregnated with photocatalyst), in collaboration with Sunstar Giken Co. We have developed TMiP (titanium mesh impregnated photocatalyst). The advantages of this photocatalytic filter are that the titanium oxide particles are heat-treated on the titanium surface to achieve high adhesion, and that, unlike conventional ceramic filters, the porous thin film of metal is highly expandable and ductile. In addition, they are lighter, stronger and less expensive than ceramic filters. In addition, they are not destroyed by plasma or ozone treatment. In short, TMiP is an ideal photocatalytic material that fulfils the two design guidelines proposed in the previous section. So far, various kinds of environmental purification systems incorporating TMiP have been studied and commercialized (Fig. 8-3). Recently, it has been reported that an air purifier combining TMiP and plasma treatment can effectively deodorize malodorous components in cigarette smoke in a smoking room, even under one-pass conditions, and that a small photocatalytic deodorizer, the "Lumineo," is sold by Maxell (Fig. 8-4). In the next chapter, Chapter 9, we will also introduce the application products of this TMiP.

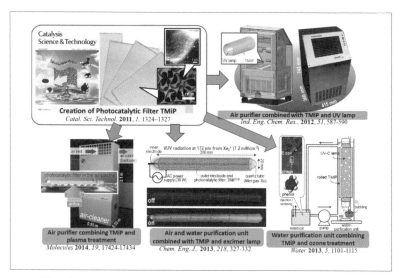

Fig. 8-3 Photocatalytic filter TMiP and its application development.

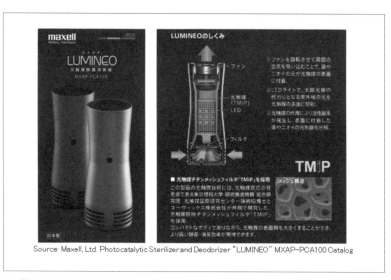

Fig. 8-4 Compact photocatalytic sterilizer and deodorizer "LUMINEO."

3 Water purification by combining photocatalytic reaction and electrolysis with boron-doped diamond electrodes

As mentioned in Section 1, efficient water purification by photocatalysis alone is difficult. Therefore, methods combining other technologies, such as ozonation, are being considered. As an example, we describe a water purification system combining electrolysis with diamond electrodes. Boron-doped diamond (BDD), which is an electronically conductive form of diamond, itself essentially an insulator, is a promising material for sensing devices and highly durable electrodes. When BDD electrodes are used as anodes, water electrolysis is known to produce powerful oxidants such as ozone and OH radicals. This means that even water highly contaminated by organic matter can be efficiently purified by direct oxidation of the organic matter on the surface of the BDD electrode and indirect oxidation by the oxidants produced. A water purification system combining electrolysis by BDD electrodes with titanium dioxide photocatalysis has been studied. The appearance and treatment flow of the system are shown in Fig. 8-5.

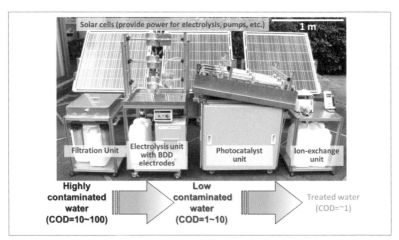

Fig. 8-5 Appearance and treatment flow of a water purification system combining electrolysis with BDD electrodes and the photocatalytic reaction.

(Ochiai, T. *et al.* Development of solar-driven electrochemical and photocatalytic water treatment system using a boron-doped diamond electrode and TiO$_2$ photocatalyst. *Water Research* 44 , 904-910 (2010))

In a preliminary test, water containing persistent substances, which are difficult to decompose by photocatalysis alone, were able to be purified efficiently by the combination with electrolysis. In order to investigate the biological purification performance of this water purification system, river water samples were treated in several steps, and the number of viable bacteria was measured before and after electrolysis and photocatalysis. In addition, the electrolysis and photocatalytic reactions during the purification process decreased the number of coliform and general bacteria to below the detection limit. Furthermore, the electricity used to drive the electrolysis and pumps during the purification process was generated by solar cells during the day. It was therefore hoped that this system could be used to secure drinking water in times of disaster.

In reality, however, the purification of river water requires filtration, electrolyte addition, ion exchange and other operations, and the system is only efficient enough to produce the equivalent of one 0.5-L PET bottle of water in a few hours. The cost of the maintenance of filters and solar cells must also be taken into account. There are still many problems to be solved for water purification by photocatalysis.

4 Conclusion

In this chapter, we have discussed one of the limitations of the photocatalytic reaction, which stems from the fact that it is a surface reaction, and design guidelines for effective environmental purification equipment based on addressing this limitation. By combining photocatalysis with other systems, as illustrated in this chapter, we can construct better environmental purification equipment by combining the advantages of each system and compensating for the disadvantages. At present, various companies and research institutes are focusing on the application of titanium dioxide photocatalysis for environmental purification to meet the needs for purification equipment due to the diversification and seriousness of environmental risks in recent years. It is important for such companies and research institutes to complement each other by sharing their know-how and specialties, as described in the design guidelines here. These efforts will lead to major breakthroughs and the construction of a comfortable and sustainable society, with the help photocatalysis.

(Tsuyoshi Ochiai)

Chapter **9**

Product Examples

**List of products registered with the photocatalysis Industry
Association of Japan**

Chapter 9

Product examples

As we have seen, titanium dioxide has two functions, "strong oxidative decomposition power" and "superhydrophilicity", and the combination of these two functions is currently used in various applications such as air purification, water purification, antibacterial action, and antifogging (Fig. 9-1). At present, the market for photocatalysis is estimated to be on the order of 100 billion yen.

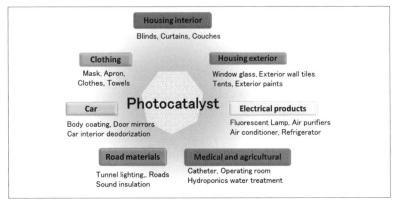

Housing interior
Blinds, Curtains, Couches

Clothing
Mask, Apron, Clothes, Towels

Housing exterior
Window glass, Exterior wall tiles
Tents, Exterior paints

Car
Body coating, Door mirrors
Car interior deodorization

Photocatalyst

Electrical products
Fluorescent Lamp, Air purifiers
Air conditioner, Refrigerator

Road materials
Tunnel lighting,. Roads
Sound insulation

Medical and agricultural
Catheter, Operating room
Hydroponics water treatment

Fig. 9-1 Applications of photocatalysis.

The first product using photocatalysis was a titanium dioxide-coated tile named "photocatalytic antibacterial, antifouling and deodorizing tile" (hereinafter referred to as "photocatalytic tile"), which was sold by TOTO (then TOTO Kiki Co., Ltd.), starting in 1993. These photocatalytic tiles are now a typical example of photocatalytic products used in everyday life.

Photocatalytic tiles have the "strong oxidative decomposition power" of the photocatalysis, which decomposes dirt and bacteria, and the "superhydrophilic effect," which means that, when rainwater hits the tiles, the water wets the tile surface underneath the dirt and washes away the attached dirt (Fig. 9-2). Thanks to these two functions, the tiles remain resistant to mold and dirt for a long time, thus reducing the cost of maintaining the house. This function has also been applied to tent materials. The white tent on the "Grand Roof" of the Yaesu Exit of Tokyo Station (Fig. 9-3) is also a photocatalytic tent, and the excellent effect of this photocatalysis makes it resistant to stains. In addition, the white tents do not require lighting during the daytime, because they are translucent, and their white color additionally helps to save energy, because they are highly reflective of sunlight and do not raise the temperature inside.

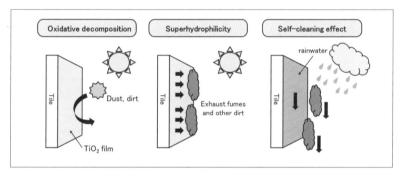

Fig. 9-2 Self-cleaning effect of photocatalysis.

Fig. 9-3 Grand roof at Yaesu exit of Tokyo station.

Furthermore, photocatalysis are being explored for medical applications such as cancer treatment, environmental purification, and agricultural applications, as well as for improving the efficiency of artificial photosynthesis, which uses solar energy directly for water splitting and the production of "solar fuels."

New developments in photocatalysis have occurred nearly every ten years since its discovery, not only in basic research on materials development and efficiency improvement, but also in applied research in the environmental and energy fields. It is expected that technologies using photocatalysis will be developed in various fields in the future, bringing about positive changes in people's lives.

In this chapter, we introduce typical examples of photocatalytic products that are applied in various fields.

1 Housing exterior: exterior tiles

Photocatalytic tiles create and maintain the beauty of houses. Today, most home builders, including Daiwa House Industry, Asahi Kasei Homes, and Ichijo Corporation, use photocatalytic technology. Panasonic Homes was founded by Konosuke Matsushita in 1963 with a strong sense of mission: "I want to make good houses that nurture the happiness of the family and provide a place for personal growth." Panasonic Homes has four key words to describe the value of the structures and technologies we use to create homes that are comfortable to live in: soft, strong, healthy, and beautiful. These are the values that apply to every home we live in. The technology employed to create the "beauty" of our homes, as a value that maintains beauty and continues to foster attachment, is photocatalytic tiles for exterior walls.

Specifically, we developed the original high-performance

photocatalytic tile "KIRATECH" together with TOTO, and now offer a rich variety of 27 colors in five patterns, in addition to the dignified and luxurious appearance of tile exterior walls (Fig. 9-4).

The photocatalytic tile "KIRATEC", which was selected for the Good Design Award in 2014, is a product with excellent functionality and design, as the iridescence phenomenon caused by the titanium dioxide composition changes the shade of the tile depending on the way the light hits it and the angle from which it is viewed, creating a unique texture. We are looking forward to seeing this product being used in all kinds of places. Other products such as Keimu's HIKARI-SERRA have also been widely used as exterior wall materials.

Fig. 9-4 Panasonic Homes photocatalytic tile
Source: Panasonic Homes Co., Ltd. HP

2   Tents for large facilities

Among the exterior building materials with photocatalytic function, tent membrane material has developed in a unique way. The roof of the Tokyo Dome, which was created in 1988, is made of tent membranes made of glass fibers coated with fluoroplastic, and is still quite strong after more than a quarter of a century. Unfortunately, however, when the Tokyo Dome was constructed, research into the application of photocatalysis was still in its infancy, so it did not have a self-cleaning function that automatically kept the surface clean. Later, Taiyo Kogyo (https://www.taiyokogyo.co.jp/), the company that made the membrane material for the Tokyo Dome, cooperated with several companies to develop the application of photocatalytic technology in tent membranes, and today, photocatalytic tents have a wide range of achievements. There are two main types of membrane materials for membrane structures that are used outdoors for long periods of time. The first is a vinyl chloride-coated membrane and the second is a fluoropolymer-coated membrane.

In the case of PVC membranes, the PVC is degraded by the oxidative degradation of titanium dioxide, so a barrier layer (protective adhesive layer) is placed in the middle and a photocatalytic layer is added on top.

On the other hand, fluoropolymers are not degraded by titanium dioxide, so fluoropolymers containing titanium dioxide particles are coated directly on top of the base fluoropolymer membrane. In this fluoropolymer membrane material, the photocatalyst particles are integrated into the membrane material and do not deteriorate, thus maintaining the photocatalytic function semi-permanently.

Originally, tent membranes were light, strong, bright inside and could be freely formed, but photocatalytic tents with a photocatalytic function on the surface have a wonderful feature that expands the possibilities of tent membranes. The basic features are the same as for other exterior building materials: (1) a self-cleaning effect that reduces maintenance costs by using the power of the sun and rain to remove dirt, (2) a clean air effect by decomposing NO_X in the air, and (3) the ability to keep the interior bright, because the surface is resistant to dirt. In addition, (4) the reflectance of sunlight is increased, so the interior is cooler and contributes to energy saving (Fig. 9-5).

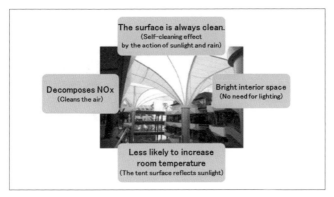

Fig. 9-5 Four features of the catalytic tent.

Source: "All About Photocatalysis Revealed by Leading Experts", by Akira Fujishima (Diamond Inc., 2017).

Taking advantage of these features, photocatalytic tented warehouses can suppress the rise in temperature in warehouses, and are now being used in warehouses for foodstuffs and medicines, which were difficult to introduce in the past. Photocatalytic tents, which can impede temperature rise and block UV rays, are also ideal for sports facilities, and are used in tennis courts, futsal grounds, batting practice domes, etc.

3   Exterior paints (coating materials)

In addition to photocatalytic tiles, paints (coating materials) are rapidly becoming the material of choice for the exterior walls of ordinary houses, high-rise buildings and factories. Ishihara Sangyo's titanium dioxide material, TOTO's Hydrotect technology, and Nippon Soda's photocatalytic coating material "Bistrator" have led this field from the beginning of the development (Fig. 9-6).

Fig. 9-6 Photocatalytic coating agent, "Bistrator."

Source: Nippon Soda Co., Ltd. HP

Starting with the development of photocatalytic tiles for interior use such as bathroom tiles, TOTO has expanded the range of its related products to include tiles for exterior walls and photocatalytic coating materials that can be applied to exterior wall materials other than tiles.

In places where high-rise buildings and factories are built, air pollution is inevitable due to chronic traffic congestion on the surrounding roads, and this directly leads to contamination of the building envelope. In Japan, tighter regulations on vehicle emissions have improved the air pollution of the high-growth period, but around the world, for example, in Beijing, the air pollution is serious, and the air pollution that occurs in parallel with the development of other cities in Asia and Africa is a global problem that has not yet been solved. When a building or factory is constructed in such a place, if the entire exterior wall of the building is covered with photocatalyst, the building itself can be kept clean by the self-cleaning function of the photocatalysis, and at the same time the air pollutant NO_X (nitrogen oxide) can be removed to keep the air clean. Cleaning the exterior walls of a high-rise building can be costly and dangerous for the workers. If we can keep the building clean while reducing the number of cleanings, we can decrease both the danger for the cleaners and the costs. An increasing number of companies are choosing photocatalytic tiles and coatings as a "trump card" to achieve this.

Recently, Japanese photocatalytic technology has been introduced to many parts of Europe and China, and more and more buildings are being coated with photocatalysts. For example, it is used on a residential building in Bremen, Germany, an apartment building in Guangdong, China, and a church building in Italy.

Composite coatings have also been developed, combining photocatalysts from Japan with materials unique to Japan. This is the case with "Eden Paint" (Fig. 9-7) of Kagoshima Eden Electric. Kagoshima is famous for Sakurajima. In fact, more than half of Kagoshima Prefecture is a "Shirasu plateau," made up of volcanic ash, pumice and other volcanic ejecta (called Shirasu). Shirasu has the same composition and structure as silica gel and alumina, which are used as adsorbents, making it a natural material that can be applied to building materials

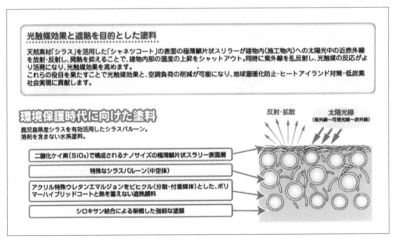

Fig. 9-7 Photocatalytic composite "Eden Paint."

Source: EDEN DENKI Co., Ltd. HP

with deodorizing and humidity-conditioning functions. It can also be processed like glass when heated to high temperatures, so it can be formed into small hollow balls (Shirasu balloons) and mixed with various materials to reduce weight and improve thermal insulation. The paint made from silica balloons is combined with our own silver-based hybrid photocatalyst, Eden Flash, to create a paint with both photocatalytic and thermal insulation properties. Combining unique materials and technologies in this way allows us to create highly original products.

4 Construction site enclosures

Temporary enclosure fences installed as a safety measure at construction sites in the city have recently been replaced by temporary enclosure fences to enhance the image of construction sites and the appearance of the cityscape. As these temporary fences are often rented, they are increasingly available with a photocatalyst coating on the surface to prevent staining. The photocatalysis not only has a self-cleaning effect but also purifies the air and deodorizes and is attracting attention as a measure for enhancing the quality of the neighborhood and the environment. For example, Nihon Kiden has developed "eco

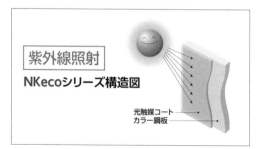
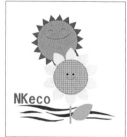

Fig. 9-8 NKeco series structure and NK eco mark.

Source: Nihon Kiden Co., Ltd. HP

photocatalytic sheet" (Fig. 9-8, Fig. 9-9), a system to stably coat photocatalyst on the surface of synthetic resin sheet, making use of their original method. These can be used by attaching them to existing temporary fences. Some of you may have already seen this product

Fig. 9-9 NKeco panel caster gate (3-m) [PCGN]. (3m) [PCGN]
Source: Nihon Kiden West Japan Sales Co., Ltd. HP

somewhere in the city as it has an NK Eco Mark sticker attached to it. If you see this mark on a construction site in town, imagine that you are using the sun's rays to clean the air, along with making sure that the surface of the temporary enclosure is kept clean.

5 | Factory exterior

Besides its self-cleaning effect, there is another reason why photocatalytic tiles are an environmentally friendly eco-technology. In addition to their self-cleaning effect, they not only break down dirt on the surface, but they also have the ability to break down and purify NO_X (nitrogen oxides), a harmful pollutant in the ambient air. In other words, the photocatalysis also has the power to clean the air by using the sun's rays. This effect is also recognized by JIS (Japanese Industrial Standards) and performance evaluation is being carried out. If we look at the natural environment, the ability to clean the air is also found in plants, such as the trees in forests. For example, the poplar tree is one of the best-known broad-leaved trees for its ability to purify the air. One manufacturer estimates that the NO_X decomposition power of 200 square meters of photocatalytic tiles is equivalent to that of about 14 poplars. The basis for this is a calculation based on the power of one poplar leaf to absorb NO_X and the number of leaves in one poplar, 15,000 (Fig. 9-10).

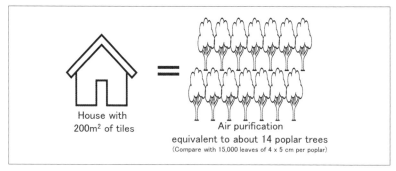

House with 200m² of tiles

Air purification equivalent to about 14 poplar trees
(Compare with 15,000 leaves of 4 x 5 cm per poplar)

Fig. 9-10 Comparing the power of photocatalysis to remove NO_X with poplar trees.

Therefore, environmentally conscious companies are actively introducing photocatalysis to the exterior walls of factories and shops. For example, Toyota Motor Corporation's factory in Toyota City, Aichi Prefecture, which produces the Prius hybrid car, uses green photocatalytic paint on the exterior walls of the building. If the air-cleaning power of the photocatalyst is converted to that of a poplar tree, the whole factory has the air-purifying power of about 2000 poplar trees. Imagine a forest of 2000 poplar trees. You cannot see it, but you can feel the workings of the photocatalysis a little closer.

6 | Window glass

Nippon Sheet Glass has commercialized photocatalytic self-cleaning tempered glass specially designed for schools (Figs. 9-11). Safety comes first in school facilities, so safety-enhancing glass such as tempered glass has been widely used. In the future, due to the importance of schools as centers for environmental education and disaster prevention, the Japanese government plans to further promote energy conservation with the aim of achieving "zero energy schools." In order to achieve the goal of zero energy consumption in schools, it is important to reduce the energy consumption of lighting, heating, cooling and ventilation. In addition to the performance of conventional toughened glass, photocatalytic self-cleaning can contribute to energy savings, and is expected to be introduced in conjunction with seismic retrofitting.

Learning about the technology used in window glazing will also help to raise environmental awareness among children. Photocatalytic self-cleaning toughened glass is being commercialized by major glass manufacturers in Japan, Europe and the USA. For example, Saint-Gobain's self-cleaning glass "BioClean" and Pilkington's "Activ" glass series are well known, and demand for these products will continue to grow.

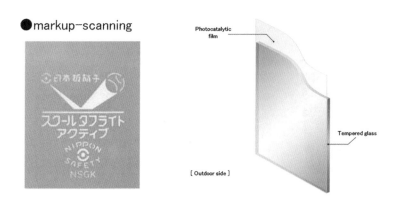

Fig. 9-11 Photocatalytic self-cleaning tempered glass for schools.

Source: Nippon Sheet Glass Co., Ltd. HP

7 Home appliances: air purifiers and air conditioners

Antibacterial, antiviral, deodorizing, and antifouling effects of interior building materials are passive functions that decompose and remove harmful substances such as bacteria, viruses, and formaldehyde that come to the surface of photocatalyst-coated materials and are effective as they are in terms of indoor air purification. In contrast, the use of photocatalysts in air purifier filters is a more active way of cleaning indoor air.

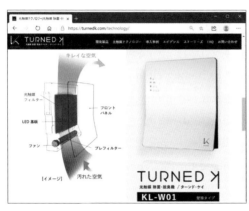

Fig. 9-12 Kaltech air purifier.

Source: Kaltech Co., Ltd. HP

Fig. 9-13 Powerful deodorizing air purifier.

Left: PLEIADES (for a 10-tatami mat room, 15.5 m²)
Right: BLUEEZE (for a 40-tatami mat room, 62 m²)

Source: Nano Wave Co., Ltd. HP

Photocatalytic air purifiers are available as general home appliances from Daikin Industries and other major companies, and, more recently, photocatalysis have been used in commercial air purifier filters that require stronger purification power, contributing to clean air in various places such as anatomy classrooms in university hospitals, pathology laboratories, nursing homes, food processing plants, pet shops, and smoking rooms in offices.

On 15 October 2020, it was reported that photocatalysis are also effective against the novel coronavirus, for which the development of anti-infection methods is currently progressing rapidly around the world. Kaltech (Fig. 9-12), which develops and sells photocatalytic air purifiers, in cooperation with RIKEN and Nihon University School of Medicine, conducted experiments on the effectiveness of a photocatalysis against the novel coronavirus (SARS-CoV-2) in an advanced facility (biosafety level-3). The effectiveness of the photocatalysis against the new coronavirus (SARS-CoV-2) was tested in an advanced facility (biosafety level-3), and the effectiveness of the photocatalysis against the infection was confirmed in a closed space.

Photocatalytic air purifiers with LED (light-emitting diode) chips, which are light-emitting elements, have also been developed. In recent years, a technology has been developed to integrate a large number of chips on a substrate to create a small and very bright light source (called "chips on board" or COB for short). Nano Wave has developed an air purifier (Figs. 9-13) that combines this technology with a photocatalysis. Inside the "Pleiades" air purifier, which measures 8.8 cm in diameter and 19.5 cm in height and is designed for a 10-tatami mat room (ca. 15.5 m^2), 156 LED chips (wavelength: 395 nm to 415 nm ultraviolet and visible light) are integrated, increasing the light output to 5,460 mW, which is about 13 times brighter than conventional LED lighting.

The structure of a photocatalytic air purifier consists of four components: (1) a pre-filter to remove dust and dirt, (2) a photocatalytic filter to decompose and remove harmful substances, odors, bacteria, and viruses, (3) a light source to trigger the photocatalytic reaction, and (4) a suction fan to circulate the air (Fig. 9-14).

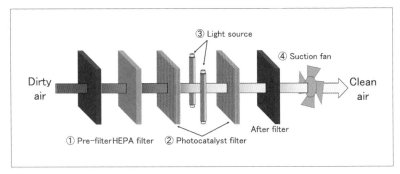

Fig. 9-14 Photocatalytic air purifier configuration.

The photocatalytic filter is designed to increase the contact efficiency with the substance to be removed by making the surface area as large as possible in order to improve air purification performance. Filter materials include honeycomb paper filters, ceramics, glass fibers and porous aluminum substrates, which are coated with photocatalyst in various ways. The photocatalyst alone acts only on the substances that come to its surface and is not effective in actively trapping the substances. The hybridization with adsorbent is one typical example, but in the development of applied products using photocatalysts, it is a major feature that the desired function is dramatically improved by combining with other technologies to realize the product's effectiveness.

In addition to black lights, mercury lamps and germicidal lamps, LEDs (light-emitting diodes) are increasingly used as a light source, making it possible to make air purifiers thinner and with lower power consumption. In the case of conventional air purifiers that are not photocatalytic, the substances trapped in the filter are not decomposed and gradually degrade the performance of the filter, while bacteria and viruses can grow on the filter and be released back into the room. With photocatalytic air purifiers, on the other hand, the decomposition and removal power of the photocatalytic reaction ensures that the performance of the filter does not deteriorate and that the indoor air is always clean.

8 | Filters

Among the functions of the photocatalytic air purifier, the focus is on the removal of airborne bacteria and viruses, and development is being carried out for higher functionality. Essential elements for high functionality of floating bacteria and virus removal are (1) efficient contact of bacteria and viruses with a photocatalytic filter, and (2) a structure that allows light to illuminate the entire surface of the filter.

As typical examples of such filters, ceramic foam and photocatalytic titanium mesh filters with a three-dimensional mesh structure have been developed. These filters have a three-dimensional random mesh structure with a large surface area, which makes them more efficient in contacting bacteria and viruses, and their many holes allow air to pass through. The filter can be washed to restore its function, even if inorganic compounds, which cannot be decomposed by Photocatalysis, are attached to the surface, and photocatalytic processing can be carried

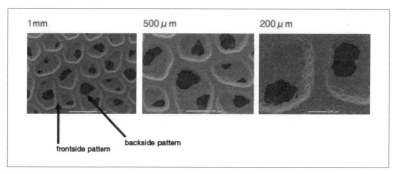

Fig. 9-15 TMiP (Titanium mesh impregnated photocatalyst).

Source: Sunstar Engineering Inc. HP

out at high temperatures to ensure stable performance.

Recently, the lightweight and flexible titanium mesh impregnated photocatalyst filter TMiP (Titanium Mesh impregnated Photocatalyst) (Fig. 9-15) developed by Sunstar Giken, an affiliate of Sunstar, has attracted attention.

The filter is made by drilling holes in a thin titanium plate in the shape of a beehive using the photoresist processing technique used in semiconductor processing, and the structure inside the holes is processed from both sides so that the air flow becomes turbulent. The perforated titanium plate is electrolytically oxidized to titanium dioxide on the surface, and titanium dioxide nanoparticles are sintered and thereby immobilized on this mesh substrate.

"TiO Clean®", an air purifier for railway cars using this TMiP, was adopted as the air purifier for railway cars on the Osaka Loop Line in 2016. The TiO Clean® air purifier has been evaluated for its ability to maintain good air quality in the cabin, and as a result, it has been gradually introduced to JR West's limited express trains from the autumn of 2020.

In addition, Sunstar will continue to develop its TMiP technology to provide a stress-free air environment that helps maintain good health ("Air that makes your body happy"). Moreover, Sunstar Giken has developed TMiP technology and commercialized the QAIS -air series (Fig. 9-16).

APS Japan K.K. has developed a photocatalytic filter based on aluminum, the "Aluminon Filter" (Fig. 9-17). This company has a high adhesion coating technology based on anodic oxidation of aluminum, which is applied to the development of photocatalytic filters. Professor Emeritus Kimiyasu Shiraki of the University of Toyama has published a paper on the effectiveness of this Aluminon filter (https://aaqr.org/articles/

aaqr-17-06-oa-0220). Based on this data, the air purifier "arc" has been developed using this aluminum-ion filter.

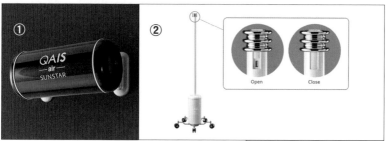

Fig. 9-16 Air deodorizer.
① QAIS -air- 01 (upper left),
② QAIS -air- 02 (upper right),
③ QAIS -air- 03 (lower right)

Source: Sunstar Inc. HP

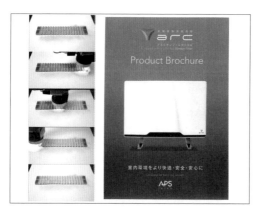

Fig. 9-17 Methylene blue degradation experiment on an aluminum ion filter (left).

(Shiraki, K.; et al., Aerosol and Air Quality Research 2017, 17, 2901-2912)

Brochure of "arc" air purifier (right).

9 | Refrigerator

A wide variety of perishable goods such as fruits and vegetables, fresh fish and flowers are transported by air on domestic and international flights. The most important factor in the transport of perishable goods is how fresh they are kept. It is known that some fruits and vegetables lose their freshness significantly due to the ethylene gas they release when they breathe.

Other products can also spoil as a result of the increase in airborne mold and bacteria, which can cause them to emit an odor. In order to reduce these problems and to keep perishable goods as fresh as possible, photocatalytic air containers (air cargo) have been developed.

JALCARGO has been using photocatalytic containers since 2004. This made it possible, for example, for highly ripe strawberries picked early in the morning in Saga Prefecture to be put on the shelves of supermarkets in Tokyo in the evening of the same day, using air cargo on the Fukuoka-Haneda flight. This is the first time in the world that photocatalytic technology has been used in air transport, and a new logo has been created to publicize the transport method. It has also led to the export of premium Japanese fruits such as strawberries, melons, cherries, peaches and apples to Asian countries with fast-growing economies.

Other examples of the application of this technology in consumer refrigerators include Hitachi Appliances, Inc. and Toshiba Lifestyle Corporation. At Hitachi Appliances, an LED light source and photocatalyst are used to break down odors in the cabinet and ethylene gas from vegetables. The carbon dioxide gas produced by this decomposition increases the concentration of carbon dioxide around the vegetables, inhibiting their respiratory activity and protecting the nutrients in the vegetables.

Toshiba Lifestyle has significantly increased its ability to remove air pollution, odors and bacteria with its new bactericidal deodorization system, which uses an Ag^+ filter and a ceramic photocatalysis filter in a more efficient structure than before.

10 Medical and agricultural sectors: hospitals

Antimicrobial tiles were the first photocatalytic products to be put to practical use, and large ceramic panels specialized for hospitals have been developed, and their application to the walls of operating rooms is expanding, as shown in Fig. 9-18.

These large photocatalytic tiles are applied to the walls of operating rooms in several hundred rooms per year. Along with its long-lasting antibacterial and antiviral effects, the tiles are resistant to scratches and stains and do not fade or deteriorate, even after the use of various disinfectants. The large size of the panels reduces the number of seams on the walls, which reduces the adhesion of bacteria; these panels have been well received by hospital staff.

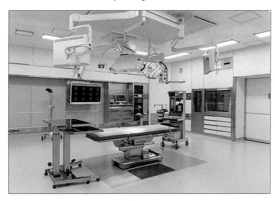

Fig. 9-18 An operating room with photocatalytic tiles on the walls.
Source: Miwa Electric Medical Co., Ltd. HP

In the future, the panels will be used not only in operating theatres, but also in intensive care units, central material rooms and any other area of the hospital where cleanliness is required and infection control is required.

11   Nursing homes

We often hear from nursing home residents that the characteristic odor is one of the serious problems in types of residence. There is a company that is trying to solve this problem by applying photocatalysis. Fujico is located in Kitakyushu City. In order to achieve deodorization that "brings smiles to the faces of users, residents and care staff alike," Fujico fully introduced deodorization and sterilization equipment to "Miyako-no-mori" (Kokurakita-ku, Kitakyushu City), a nursing home complex that includes a small-scale multi-functional care facility, day service, residential nursing home and group home. At the time of construction, a photocatalytic product called Mask Shield (deodorizing and sterilizing tiles) (Fig. 9-20), which is based on Fujikoh's unique thermal spraying technology (Fig. 9-19), was applied over the entire two-story facility. Since the opening of the facility in April 2012, the deodorizing effect of the tiles has continued to this day. The nurses who visited Miyako-no-Mori for the first time were surprised to find that there was no odor. It seems that Fujico's photocatalytic products are very useful for users, residents and their families to live comfortably and healthily, and to provide a comfortable and stress-free workplace.

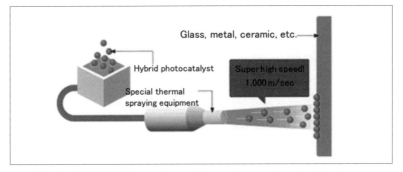

Fig. 9-19 FUJICO's unique thermal spraying technology.

Source: FUJICO CO., LTD. HP

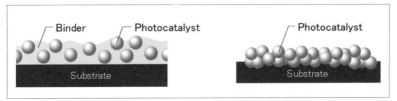

Fig. 9-20 Mask shield (deodorant and disinfectant tiles).

Source: FUJICO CO., LTD. HP

12 | Water purification system

Photocatalytic systems are used in some cases for water treatment of a limited amount of water, not water from rivers. For example, equipment incorporating photocatalytic fibers has been developed to purify water above ground, and has been successful in preventing legionella bacteria in thermal baths. The photocatalytic fiber, which acts as a filter, has a felt-like structure that efficiently contacts and breaks down bacteria and pollutants in the water in three dimensions. Hot springs and bathing facilities are being built all over the country as a service to local residents and as a feature of local tourism, but there have been a number of outbreaks of Legionella bacteria and other problems, and keeping the hot water and hot spring water in public baths clean has become an issue. In particular, in hot springs with high alkalinity, it is difficult to achieve the desired effect with ordinary chlorine disinfection alone. However, photocatalytic water purifiers have been found to be highly effective. The best feature of photocatalytic systems is that they can decompose even the dead bodies of bacteria, so that dirty water can be recycled into clean water (Fig. 9-21).

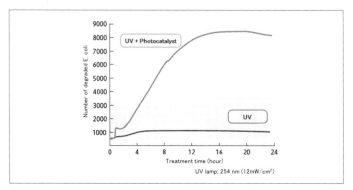

Fig. 9-21 Degradation of escherichia coli by photocatalytic fibers (compared to UV light only).

Source: "All About Photocatalysis Revealed by Leading Experts", by Akira Fujishima (Diamond Inc., 2017).

Photocatalytic water purification systems are not specialized for sterilization but are also effective in general industrial applications such as the decomposition of persistent and poisonous dioxins, PCBs, and cyanide compounds. For example, it has been proven to break down almost 100 % of dioxins in water.

13 Purification of groundwater

Several methods have been proposed for decomposing and treating volatile organochlorine compounds such as tetrachloroethylene (also known as perchloroethylene, PCE), which is used as a solvent in dry cleaning, but these methods have not been widely used due to the problems of generating toxic substances.

The photocatalytic method has been developed as a promising method because it can decompose and detoxify PCE, etc. Let us introduce the soil groundwater purification system developed by ADEKA Engineering & Construction Co., Ltd. (Fig. 9-22).

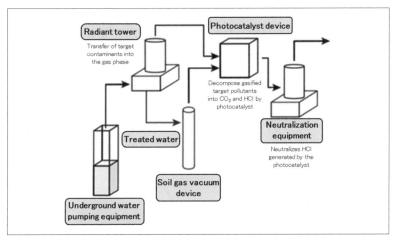

Fig. 9-22 Soil and groundwater purification system using photocatalysis.

Source: "All About Photocatalysis Revealed by Leading Experts", by Akira Fujishima (Diamond Inc., 2017).

Volatile organochlorine compounds collected from soil and groundwater are gasified and passed through a photocatalytic system to be decomposed into carbon dioxide, water and hydrogen chloride. The photocatalytic system consists of a light source and a number of reactors with titanium dioxide fixed on a ceramic surface. The decomposition products of the photocatalytic reaction are then neutralized with alkali in a neutralization unit to render them harmless, and they are finally released into the atmosphere. More than 70 of these systems are already in use for the purification of soil and groundwater, and in recent years they have been enlarged to increase their capacity.

During the purification of soil and groundwater, pollutants are evaporated and treated as a gas by the photocatalytic filter. In other words, it is a photocatalytic treatment of gaseous substances in the air, which is the same as the function of an air purifier.

Essentially, the treatment of pollutants in water is a very difficult subject, first, because the small amounts of pollutants in the water have to be directed to the photocatalytic filter, and the filter has to be well illuminated. Furthermore, the limited solubility of O_2. is why photocatalytic treatment of trace pollutants in water is difficult. Of course, there is also a way to use titanium dioxide particles in suspension (individual particles dispersed in a liquid), but, in this case, the separation of the titanium dioxide particles in the water is a major problem. As one way to solve this problem, a young Indian researcher is using a method of coating titanium dioxide on iron particles to create a photocatalytic system (i.e. a photocatalytic reaction system) and then using magnets to collect the titanium dioxide particles after photocatalysis. This is quite an interesting idea.

14 Road materials: tunnel lighting

The Jōshin'etsu Expressway was built in time for the 1998 Nagano Olympics, and, for the first time, photocatalytic tunnel lighting equipment was used. We don't usually think about lighting when we pass through motorway tunnels, but we can all easily recall the experience of being stuck in a traffic jam due to lane restrictions near a tunnel. The problem is that car exhaust contains not only NO_X, but also oil and carbon, so the light fittings in tunnels are easily soot-blackened and reduced in brightness. Dark tunnels increase the risk of accidents, so we clean them regularly to remove this soot. In other words, for safety reasons, the light fittings in the tunnels needed to be cleaned regularly, but this requires lane restrictions, causing traffic congestion and putting the cleaners at risk.

One of the measures to alleviate this situation was to use photocatalytic self-cleaning cover glass for lighting equipment in tunnels, which was proposed mainly by the Japan Highway Public Corporation, and after demonstration tests, was realized during the Nagano Olympics. In 1996, the Illuminating Engineering Institute of Japan (IESJ) awarded Fujishima and his research group the "Japan Lighting Award" for this achievement. At the beginning of the application research, some people said that it was just a technology to reduce dirt by oxidative decomposition, or that it was a technology that did not need to exist in the world, but the fact that it turned out to be useful, even if only a little, for avoiding traffic jams on highways and improving the safety of workers motivated the research group even more.

The compatibility between lighting equipment, which is intrinsically a light source, and the photocatalyst, which necessarily needs light for its reaction, was proven under the demanding conditions of a motorway tunnel, and the anti-staining effect of the lighting equipment was subsequently applied to outdoor lighting equipment around roads, such as road and street lights.

15 | Sound insulation walls

The photocatalytic reaction effect can be used in various materials around roads by utilizing the natural energy of sunlight and rainfall.

One example is the polycarbonate transparent sound barriers installed on motorways. The transparency provides visibility and reduces the feeling of oppression, both for the occupants of the vehicle and for the surrounding population. However, we have heard that there are many places where the transparency is soiled by exhaust fumes and other pollutants, thus spoiling the view.

Photocatalytic coating has a self-cleaning effect and helps to maintain visibility in these areas. In the same way, the application of photocatalytic coating to road signs, eye guides, road reflectors, and roadside signs is also progressing (Fig. 9-23).

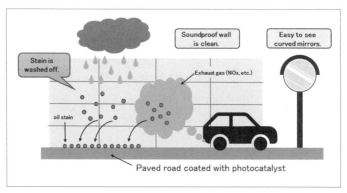

Fig. 9-23 Various applications on road surfaces and roadside.

Specific products include anti-fogging and anti-drip curve mirrors, Hydroclean mirrors, anti-fogging, anti-drip and anti-fouling translucent sound barriers, and photocatalytic superhydrophilic transparent sheets (Hydroclean transparent sheets) from Sekisui Plastics (Fig. 9-24).

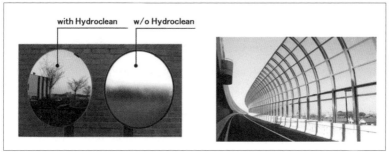

**Fig. 9-24 Hydroclean mirror (left),
Hydroclean transparent plate (right).**

Source: Sekisui Jushi Corporation HP

16 | Roads

Did you know that paving technology has become more sophisticated in recent years? You may have noticed that on a rainy day, when driving on a motorway, there are places where it is difficult to see the road ahead because of the spray of water, while on other sections of the road, visibility is good, because there is no accumulation of rainwater on the surface and no spray of water. This is what is known as high performance paving, where the surface layer of the pavement has more clearance than usual, in order to improve drainage. This means that rainfall will soak into these gaps rather than pooling on the road surface, greatly reducing the amount of spray from vehicles and maintaining good visibility.

The benefits of high-performance paving do not stop there. The noise emitted by a car on the road can have a negative impact on the environment of the surrounding area, the more cars there are, the more noise is absorbed and reduced by the high performance pavement sections, which also have environmental benefits.

The "Photo-Road Method," which uses photocatalysis to clean the air on the road surface, is a further upgrade of this excellent technology. By applying a photocatalytic coating to the surface of a high-performance pavement with high drainage and many gaps using a special cement-based immobilizing material, it is possible to decompose and remove pollutants such as nitrogen oxides (NO_X) contained in car exhaust gases. The idea was to remove the exhaust gases from the car before they spread widely in the atmosphere, right on the road surface. It was a long time ago, but Fujita Road, the company that developed the system, won the Nikkei BP Technology Award (2004) and the first Eco Products Award (2004).

NO_X is a problematic air pollutant, including nitric oxide (NO) and nitrogen dioxide (NO_2). In the constant monitoring of air pollution, the sum of both is monitored. The NO_X emitted by cars is oxidized by the photocatalyst coating on the road surface and by the action of the sun and is then fixed to the road surface as neutral calcium nitrate, which compounds with calcium, the main component of the photocatalyst fixative. When it rains, the calcium nitrate is washed away as harmless nitrate and calcium ions, and the road surface is restored to its original state.

The greatest advantage of the "Photo-Road Method" is that once the installation is completed, the only energy source for cleaning the air is the natural power of sunlight and rainfall, just as in the case of the photocatalytic tiles for exterior use, and there is no need for power or special maintenance. A further benefit is that the photocatalytic coating on the road surface improves the wear resistance of the road.

The "Photo-Road Method" is an original Japanese technology developed jointly by companies in Japan, and has been used in numerous projects, such as on Loop 7, Meiji-dori (Tokyo), National

Routes 14 and 16, Prefectural Road Ichikawa-Urayasu (Chiba), in Saitama New City Centre, on the approach to Hikawa Shrine (Saitama) and in the Motohama green area in Amagasaki (Hyogo).

Furthermore, overseas, demonstration tests have been carried out in cities such as Paris (France), Bergamo (Italy) and Antwerp (Belgium), with positive results. Examples in China and Korea will be explained in later chapters.

17 Car-related: side mirrors

Photocatalysis are used in many ways on and around roads, but they are also being used in cars. One application is in door mirrors. On rainy days, when door mirrors are fogged up and difficult to see, this is a problem that needs to be solved, from the point of view of both safety and driving comfort. The superhydrophilic effect of the photocatalysis prevents water droplets from forming and spreading, thus ensuring good visibility without fogging up the mirrors.

Furthermore, the addition of silica, which is relatively hydrophilic even in dark areas, has been designed to maintain the anti-fogging effect even when there is no light. It is now a standard feature in Toyota's luxury cars and is beginning to be used in other manufacturers' standard cars. The self-cleaning effect, which is widely applied in the construction sector, is also beginning to be used in the field of car body coatings as a dirt-resistant coating method. However, the application of photocatalytic coatings on car bodies should start with buses and trucks, as there are many issues that need to be resolved before they can be applied for private cars.

Various materials are used for photocatalytic filters, which are the heart of photocatalytic air purifiers, but ceramic materials are the most widespread for commercial filters. Ceramic photocatalytic filters are designed so that they can be used again after cleaning even if their performance deteriorates, and running costs can be reduced for long-term use. As a result, they have been installed in a great variety of locations. They have been installed in hospitals, welfare facilities, hotels, restaurants, offices and warehouses where the odor concentration is relatively low.

They are also being installed in places with high concentrations of malodorous compounds, such as animal testing facilities in research laboratories, food processing plants, waste treatment plants, animal barns and animal manure treatment facilities. The reason why our products are used in such a wide variety of locations is that we have designed filters and other equipment to suit the concentration and occurrence of the malodorous substances to be treated.

While the photocatalytic air purifier developed jointly with Seiwa Industries (now Seiwa Environmental Engineering Co., Ltd.) was being used in various places, it was also introduced to the Tokaido and Sanyo Shinkansen bullet trains with the cooperation of Andes Electric (Fig. 9-25).

The Tokaido and Sanyo Shinkansen trains are now fully non-smoking, and smoking booths have been installed on the decks, but if smoke and odors from the smoking booths leak into the train, it would be meaningless to make the trains fully non-smoking. It was therefore of paramount importance to remove as much tobacco smoke and odor as possible from smoking areas. Against the backdrop of an era in which

Fig. 9-25 Shinkansen smoking room (left), The filter part of the air purifier in the smoking room of a Shinkansen (right).

Source: "All About Photocatalysis Revealed by Leading Experts", by Akira Fujishima (Diamond Inc., 2017).

the effects of passive smoking on the human body have been highlighted as a health issue, there has been a growing awareness of the need to separate smoking areas. In addition, it is not only a matter of getting to the destination quickly, but also of having a healthy and comfortable travel space, which is one of the main reasons for the introduction of high-performance air purifiers. Conventional measures to separate tobacco smoke, such as electrostatic precipitators, remove visible smoke, but not ammonia or aldehydes contained in the smoke.

In contrast, photocatalytic systems can decompose and remove even these gaseous components, which is why they are used in the smoking rooms of the Shinkansen. On the Shinkansen, you don't usually notice it at all, but on the surface of the filter of the environmental purification system hidden in the ceiling of the smoking room, a photocatalytic reaction takes place and works to keep the air clean.

19 Clothing: masks

Fujitsu has developed and commercialized a high-performance

Fig. 9-26 Anti-Hay fever mask.

Source: FUJITSU JOURNAL HP

photocatalytic material called titanium apatite. This material is made by introducing titanium ions into calcium hydroxyapatite, which is found in human bones and teeth.

Calcium hydroxyapatite has the property of easy adsorption, and the photocatalysis has the property of decomposing bacteria, viruses and organic matter.

As shown in Fig. 9-26, it is commercially available as an anti-hay fever mask because of its excellent adsorption.

20 Aprons

Fig. 9-27 Photocatalytic apron (Kireina apron).

Source: ALT corporation HP

Gaia developed Gaia Clean (GCT series), a one-component photocatalytic processing liquid containing a binder that hardens at room temperature, for processing on textile and organic materials such as clothes, curtains, and carpets.

It is processed and sold for aprons (Fig. 9-27), work clothes to be used under high perspiration conditions, sportswear and suits.

21 Fabric products

Fig. 9-28 Photocatalytic spray MX-AZ03JK

Source: Sharp Corporation HP

There have been many applications of photocatalysis to clothing. For example, Atugi has applied photocatalysis to stockings and tights and has commercialized them.

An increasing number of companies are also selling spray-type products. As an example of a photocatalytic spray, Sharp launched on 17 July 2020 a "photocatalytic spray" (Fig. 9-28) that uses a unique visible light-responsive photocatalyst that reacts not only to sunlight but also to light from indoor lighting to produce a highly effective deodorizing, antibacterial and antiviral effect.

This spray employs a unique visible light-responsive photocatalyst, which is the result of applying the opto-semiconductor technology and powder processing technology that Sharp has cultivated through the development of multifunction devices. Because it reacts to light in a longer wavelength range than titanium dioxide, it has excellent decomposition capabilities, even in the light of indoor lighting. It can be sprayed on wallpaper, floors, furniture and even clothing.

22 Towels

Aska sells a towel "Saratto Dry®" (Fig. 9-29), which is loaded with a photocatalyst in the structure of the microfiber fibers and has a strong antibacterial and deodorizing effect when dried in the sun. The Japan Spinners' Inspection Association has confirmed that the antibacterial and deodorizing effects of the photocatalysis in this towel remain stable over a long period of time, even after 50 washes.

Although there are products that use photocatalysts on cotton towels, the effect and sustainability are not as good as Saratto Dry.

Fig. 9-29 Photocatalytic towel "Saratto Dry®".
Source: Aska CO.,LTD. HP

23 | Home interior: blinds and curtains

Photocatalytic products are spreading in relation to wall and window surfaces for interior decoration. For example, anti-viral curtains have been launched that reduce the risk of viral infection by applying a photocatalytic treatment to the surface. When light hits the surface of the curtain, oxygen and moisture in the air cause a chemical reaction on the photocatalyst, and the radicals produced attack and destroy the virus. The photocatalytic material itself does not change during this process, so the effect lasts for a long time, and the titanium dioxide does not desorb when washed.

As a wallpaper, Tosa Washi wallpaper "Tosalite" is available with photocatalytic treatment. By applying a photocatalytic treatment to the washi (Japanese paper), it has the ability to decompose and remove airborne bacteria and harmful substances, as well as removing tobacco and pet odors, making it a high value-added wallpaper. This is a unique product that combines the traditional Japanese paper-making techniques with the scientific technology of photocatalysis, and we believe that it will be increasingly used in hotels and accommodation facilities both in Japan and overseas, as well as in general housing.

As for photocatalytic products related to windows, blinds have also been developed from an early stage. By applying a photocatalytic coating to the surface of the blinds' blades, organic stains such as oil stains can be broken down and removed. It also has antibacterial, deodorizing and mold-inhibiting properties. Nichibei, TACHIKAWA Blinds, TOSO and others have commercialized this product. (Figure 9-30).

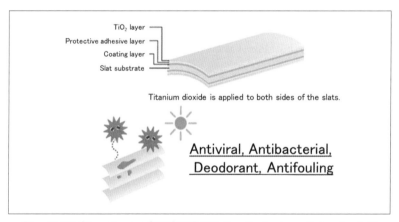

Fig. 9-30 Structure of NICHIBEI's [anti-viral, anti-bacterial] titanium dioxide coated thermal barrier slats.
Source: NICHIBEI CO., LTD. HP

24 | Lighting

Since the discovery of the photocatalytic effect, there have been many studies on the application and development of lighting-related products,

Fig. 9-31 Deodorizing LED bulbs.

Source: Kaltech Co., Ltd. HP

due to the intrinsic availability of light. At Toshiba Lighting and Technology Corporation, the application to fluorescent lamps was studied, and a new fluorescent lamp with photocatalyst was developed. The main purpose was to break down dirt in lighting equipment. As an example of application, it was used for study stands, etc., with the expecting of decomposing formaldehyde, which is considered to be one of the causes of the "sick buiding" syndrome.

On the other hand, there has been a recent switch from fluorescent to LED lighting in homes, offices, factories and many other places. This has led to the development and application of photocatalytic thin film coatings for LED lighting. As an example, Kaltech has commercialized a deodorizing LED light bulb (Fig. 9-31), which combines LED lighting with a photocatalytic deodorizer.

In addition, advances in LED manufacturing technology have led to the use of LEDs capable of emitting deep ultraviolet light, which has a shorter wavelength (200-300 nm). This deep UV radiation can excite photocatalysts and also has the effect of denaturing biological DNA. Therefore, it can be applied to products with a high antibacterial and antiviral effect. As an example, the product of Taiyo Kogyo (https://www.taiyo-technologies.jp/) is shown in Figure 9-32 (This Taiyo Kogyo, is different from Taiyo Kogyo, which is famous for large facility tents). This product uses a combination of deep UV LEDs with a wavelength of 275 nm and UV LEDs with a wavelength of 365 nm to further enhance the effect.

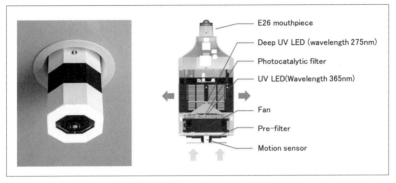

E26 mouthpiece

Deep UV LED (wavelength 275nm)

Photocatalytic filter

UV LED(Wavelength 365nm)

Fan

Pre-filter

Motion sensor

Fig. 9-32 LED deodorization lighting.

Publisher: TAIYO KOGYO CO., LTD. HP

25 | Photocatalytic mosquito repellent

We have developed a photocatalytic mosquito repellent in collaboration with Earth Chemical and Sunstar Giken, which uses a photocatalyst and no insecticides. The mosquito repellent uses carbon dioxide generated by the photocatalysis to lure mosquitoes in, and then turns a fan to capture the mosquitoes (Fig. 9-33).

In addition to domestic use, we hope to install the system in countries in Africa and Southeast Asia that are suffering from infectious diseases such as malaria transmitted by mosquitoes. In recent years, mosquitoes infected with dengue fever have become a problem in Japan. It is the blood-sucking mosquitoes (females) that transmit infectious diseases such as dengue fever and malaria. The main attractants known to lure blood-sucking mosquitoes are (1) carbon dioxide, (2) odor (lactic acid and other attractants), (3) warmth, and (4) color.

Some mosquito repellents use titanium dioxide and ultraviolet light, and the photocatalytic effect (oxidative decomposition) breaks down dirt

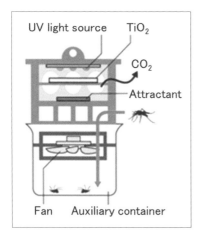

Fig. 9-33 Schematic of a photocatalytic mosquito repellent.

Source: "All About Photocatalysis Revealed by Leading Experts", by Akira Fujishima (Diamond Inc., 2017).

on the titanium dioxide into carbon dioxide, and the synergistic effect of the ultraviolet light attracts mosquitoes and sucks them away. However, the amount of carbon dioxide gas produced is very small, so it is difficult to achieve the desired effect. There are also mosquito traps on the market that use carbon dioxide gas cylinders, but they are too heavy to move around and the cylinders are difficult to change. Fujishima's group developed a mosquito repellent using titanium dioxide photocatalysis sheets, which have high efficiency in decomposing organic gases (Fig. 9-34). By applying the newly developed titanium dioxide photocatalytic sheet to the mosquito repellent, the efficiency of decomposition of organic gas, i.e. generation of carbon dioxide gas, was increased, and the efficiency of mosquito collection was successfully improved compared with similar products.

Fig. 9-34 Earth mosquito-sweeper.

List of products registered with the Photocatalysis Industry Association of Japan

As detailed in Chapter 14, the Photocatalysis Industry Association of Japan was established on 1 April 2006 by a group of companies dealing with photocatalysis. The Photocatalysis Industry Association of Japan formulates product standards and operates certification marks for the proper dissemination of photocatalytic products, promotes understanding of photocatalytic products to consumers, establishes international standards, and conducts publicity activities.

The website of the Photocatalysis Industry Association of Japan (PIAJ) provides various types of up-to-date information on photocatalysis, such as information on the products of member companies, PIAJ-certified products, registered materials and general photocatalysis information. As one of these items, the list of products registered with the Photocatalysis Industry Association of Japan is introduced below. For details, please refer to the following website.

List of products registered with the Photocatalysis Industry Association of Japan
https://www.piaj.gr.jp/en/piaj-mark-en/catalogue-en/

(Katsunori Tsunoda, Tsuyoshi Ochiai)

Chapter **10**

Antibacterial and Antiviral Performance Evaluation Method

Photocatalytic antibacterial and antiviral mechanisms and their usefulness

Evaluation of anti-microbial activity by the JIS/ISO method

Photocatalytic antibacterial and antiviral mechanisms and their usefulness

Microorganisms and viruses are organic substances composed of proteins and nucleic acids, and are very small organisms (Table 10-1). Therefore, photocatalytic reactions can degrade microorganisms and viruses, causing them to lose their ability to infect and multiply. This is the antimicrobial-antiviral mechanism of photocatalysis. In this way, photocatalysts can reduce the risk of microbial infection. In addition, it is possible to prevent the development of resistant micro-organisms and viruses that can occur with the use of common antibacterial and antiviral agents. For example, methicillin-resistant Staphylococcus aureus is a major threat to people's lives as it can cause a hospital-

Table 10-1. Comparison of bacteria and viruses.

	bacteria		virus		
Target	Gram-negative (*E. coli, K. pneumoniae*, etc.)	Gram-positive (*S. aureus*, etc.)	Bacterio-phage Qβ	Feline calicivirus	Influenza virus
Size	3 μm	1 μm	20 nm	30 nm	100 nm
Multiplication	Self-propagation		No self-propagation (Host: bacteria)	No self-propagation (Host: cells)	
Characteristics of the structure	Lipopolysaccharide, Outer, membrane, Thin peptidoglycan layer, Inner membrane	Thick peptidoglycan layer Cell membrane	No envelope	No envelope	With envelope

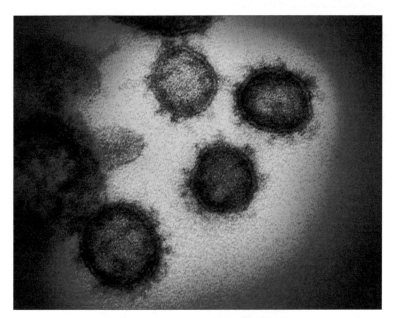

Photomicrograph of SARS-CoV-2.

Coronavirus covid-19
Source: Pacific Press Service, Photographer: IMAGE POINT FR-LPN/BSIP, Date: Feb. 2020

acquired infection (healthcare-associated infection). A photocatalyst is a material that can reduce the risk of infection while suppressing its occurrence, and many products are being developed. On the other hand, it is important to use a standard test method for evaluating their performances. For this reason, standard test methods such as JIS/ISO have been established and utilized, as described below.

Evaluation of anti-microbial activity by the JIS/ISO method

Table 10-2 shows the list of standard test methods for anti-microbial activity by photocatalysis. As shown here, JIS/ISO tests have been established for antibacterial, antiviral, antifungal and algaecide, and they are used for the development of various materials and products. These evaluation methods are periodically reviewed to ensure that they remain relevant and easy to use. In this Section, we introduce the basic antimicrobial performance evaluation methods the commonly used and introduce the differences between them and the test methods for other microorganisms. It is important to note that some microorganisms may have the risk of infection to a technician, so it is essential that the test is carried out by a person who is familiar with biosafety and testing techniques. Biosafety-grade facilities and equipment are also required. Safety cabinets, autoclaves, gloves, etc. should be available. (Fig. 10-1, Fig. 10-2)

Table 10-2. List of JIS/ISO tests.

Subject	Light source	JIS	ISO
Bacteria	UV	JIS R 1702	ISO 27447
	Vis.	JIS R 1752	ISO 17094
		—	ISO 22551
Virus	UV	JIS R 1706	ISO 18061
	Vis.	JIS R 1756	ISO 18071
Algae	UV	—	ISO 19635
Fungus	Vis.	JIS R 1705	ISO 13125

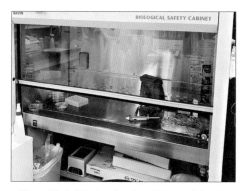

Fig. 10-1 Biological safety cabinet.

In this cabinet, technicians can handle infectious microorganisms while being protecting from infection.

Fig. 10-2 Autoclave.

Consumables and test pieces used in the test are first disinfected in the autoclave.

1　Antimicrobial performance evaluation method

Antimicrobial performance evaluation methods can be divided into two main categories: flat products such as glass and tiles, and fibrous products. For flat products, a method called the film adhesion method is used, and for fibrous products, a method called the glass adhesion method is used. The bacteria used in the two methods are different: *E. coli* and *S. aureus* are used in the film adhesion method, while *E. coli* and *K. pneumoniae* are used in the glass adhesion method. Bacteria have two main characteristics (Gram-negative and Gram-positive bacteria). Therefore, two types of bacteria are used in the performance evaluation: Gram-negative bacteria (*E. coli* and *K. pneumoniae*) and Gram-positive bacteria (*S. aureus*). The basic procedure is as follows.

Film adhesion method

① 　Cultivation of the test bacteria solution

Bacteria for the test are cultured on agar and mixed into the solution to ensure that the number of bacteria is within the test range. At this time, the bacteria must be properly controlled.

② Preparation of the test article

Prepare 6 photocatalyst-treated and 9 non-treated test pieces of 50 mm square size. All test pieces are prepared in such a way that they are free from contamination by micro-organisms other than the test organisms and the surface is free from organic contamination. If there is contamination from other microorganisms or organic matter, it is not possible to properly check the antibacterial effect of the photocatalytic reaction.

③ Installation of the test product

Set up the test product and test solution as shown in Fig. 10-3. Figure 10-4 shows an example of the process. The amount of test solution can be changed according to the size and characteristics of the test product.

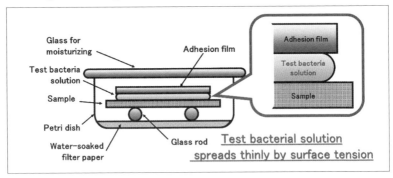

Fig. 10-3. Installation method of test specimen (film adhesion method).

Fig. 10-4. Inoculation of bacterial solution and installation of film.

The test piece is inoculated with the test solution (left) and then covered with the adhesion film (right).

④ Photocatalytic reaction

The photocatalytic reaction is carried out at a maximum UV light intensity of 0.25 mW/cm^2 for a certain time. An example of the matrix in the performance evaluation test is shown in Table 10-3.

Table 10-3. Test matrix (example of ultraviolet light antimicrobial test).

Time	0 hr	8 hr	
Light intensity	UV 0.0 mW/cm^2	UV 0.0 mW/cm^2	UV 0.1 mW/cm^2
w/o photocatalyst	#1 #2 #3	#4 #5 #6	#10 #11 #12
with photocatalyst	—	#7 #8 #9	#13 #14 #15

As shown in Table 10-3, the test is divided into three groups: a group at time 0 (immediately after inoculation), a group without UV light irradiation and a group with UV light irradiation. If visible light-

Fig. 10-5. Example of light irradiation (visible light).
Irradiation is carried out in the box.

responsive photocatalysts are used instead of UV light, an irradiance of up to 3000 lx is used, and a filter is placed between the specimen and the light source to cut off the small amount of UV light emitted by fluorescent lamps. Whichever light source is used, the irradiance should be the one that is expected to be used in practice. Figure 10-5 shows an example of light irradiation.

⑤　Confirmation of viable bacteria count

After the photocatalytic reaction, the inoculated test solution is collected, and the test solution is diluted 10 times and cultured. By culturing, viable bacteria grow in the medium and form colonies (Fig. 10-6), so it is possible to visually check the bacteria number. The number of viable bacteria is measured (Fig. 10-7).

⑥　Confirmation of the antibacterial effect

The test conditions are confirmed from the values of the unprocessed product in the viable bacteria count obtained. Once the test is established, the antimicrobial effect is calculated. The antimicrobial effect is calculated as the antimicrobial activity value based on the difference in decrease between the untreated and treated products after light irradiation. The photocatalytic effect can be checked by comparing the untreated and treated products without light irradiation and subtracting the value obtained from the antimicrobial activity value. The antimicrobial effect is judged to be present when the antimicrobial activity value is 2.0 or higher (it depends on the criteria at each laboratory).

Glass adhesion method

① **Cultivation of bacterial solution for testing**

The bacteria solution used in the glass adhesion method is cultivated on agar media, and then the bacteria is cultivated in liquid form the day before the test. The bacteria are then cultured in liquid form on the day before the test, and again on the day of the test to ensure that the number of bacteria remains within the specified limits.

② **Preparation of the test piece**

The basic size of the test pieces is 50 mm square. Six pieces are prepared with photocatalytic treatment and 9 pieces without photocatalytic treatment. The test pieces are sterilized by high-pressure steam in an autoclave to ensure that there is no contamination by micro-organisms other than the test bacteria. In order to remove organic contamination on the surface of the test article beforehand, it may be used after being irradiated with ultraviolet light, as in the film adhesion method.

③ **Setting up the test article**

The test article is placed in a Petri dish moistened with water and a U-tube, on top of which a 60 mm square glass plate is placed. The test piece is then inoculated with a fixed amount of the bacterial solution. After inoculation, the glass plate is placed between the inoculum and the glass to retain moisture, and the photocatalytic reaction takes place. After the photocatalytic reaction, the method is the same as the film adhesion method until the viable bacteria count is confirmed.

④ **Confirmation of antibacterial effect**

Of the viable bacteria count obtained, the test conditions are confirmed from the value of the untreated product, and the antibacterial effect is confirmed in the same way as the film adhesion method.

In this way, the procedure of the antimicrobial test is very simple and the procedure is common to other microorganisms as much as possible. The following page shows a review of the general flow of antimicrobial performance evaluation testing. Even if "antimicrobial" here is replaced by other microorganisms, the test will be conducted in accordance with this flow, so we hope you will keep this in mind.

● **Review of Antimicrobial Performance Evaluation Tests**

Antimicrobial performance evaluation and other anti-microbial performance evaluation tests proceed as follows.

Select the test method according to the nature and shape of the photocatalytic product to be tested.

Select whether the JIS/ISO standard can be used or whether the test should be applied. If applied testing is necessary, please consult with a research institute such as KISTEC.

↓

Selecting and obtaining the bacteria to be tested

Select, obtain, and increase the number of bacteria that can be used for the purpose.

↓

Preparation, installation, and photocatalytic reaction of test products

Preparation, installation, and photocatalytic reaction of the test article according to the selected standard.

↓

Confirmation of antibacterial effect

Check the obtained colonies, etc. and calculate the antimicrobial effect.

To reduce the risk of infection, the work should be done in a safety cabinet, and the microorganisms should be sterilized in an autoclave as shown in Figure 10-2 as soon as possible after the test is completed. It is also a good idea to keep a spray or bottle of disinfectant readily available at hand.

2 Antimicrobial performance evaluation method assuming real environment

The antimicrobial performance evaluation method specified by JIS/ISO is to inoculate the test product with a bacterial solution, collect it after the photocatalytic reaction, and measure the number of bacteria. On the other hand, it can be said that in our living environment, there is hardly any situation where there is a liquid containing many bacteria as in the JIS/ISO antimicrobial test method. In fact, as a result of verifying the antimicrobial effect of photocatalytic products in the real environment for one year, we found that the antimicrobial effect was lower than the result of the JIS/ISO evaluation. Therefore, a method for evaluating antimicrobial performance in actual environments was developed and established as ISO 22551. The key point of this method is the inoculation method of the test solution. Unlike the usual JIS/ISO method, a viscous test solution is prepared and directly applied to the test product to measure the antimicrobial effect, but no filter paper is added to simulate a living environment to confirm the antimicrobial effect in a living environment. As a result of actual testing using this method, results similar to the antimicrobial effect obtained in a real environment can be obtained in the laboratory.

3 Antiviral performance evaluation method

In the case of non-photocatalytic processed products, influenza virus and feline calicivirus (used as an alternative to norovirus) are mainly used, but the target virus used in the photocatalytic antiviral performance evaluation method is a type of virus called a bacteriophage. Bacteriophages are viruses that do not infect animal cells like influenza virus or feline calicivirus, but rather infect bacteria and multiply. The advantages of using bacteriophages are as follows. Firstly, the cost of the test is low. The cost of culturing and maintaining animal cells is much higher than the cost of culturing and maintaining bacteria. Secondly, there is the safety factor. Bacteriophages are not infectious, so they can be safely tested on human subjects. Also, as mentioned above, the antiviral effect of the photocatalysis is due to the degradation reaction, which means that bacteriophages can be used as a surrogate model for real viruses. For this reason, bacteriophages are used in the antiviral evaluation of photocatalysis.

The difference with the antimicrobial performance evaluation method is in the multiplication and detection of the bacteriophages. In order to increase the number of bacteriophages, a host bacterium is grown beforehand and then infected with bacteriophages. When the bacteriophage is further cultured in the infected state, the number of infected bacteriophages increases. The bacteriophage can then be collected to produce a bacteriophage solution. The bacteriophage solution is then prepared so that it is within the test range. From there, it

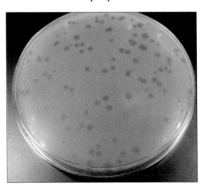

Fig. 10-8 Detection of bacteriophages.

Each transparent particle is a plaque that has grown from a single bacteriophage.

is collected in the same way as for the antimicrobial test until it is infected with a host bacterium, which is then spread on an agar medium and cultured. The infectious bacteriophage then dissolves the host bacteria, so that plaques become visible (Fig. 10-8). This allows the amount of bacteriophage to be measured and the antiviral effect to be confirmed. In addition, using the JIS/ISO as a reference, it is possible to conduct tests using viruses that infect actual animal cells (e.g., influenza virus and feline calicivirus) to confirm the antiviral effect more concretely, which is considered to be important for research and development. Fig. 10-9 shows the infection titer after the virus has been increased. As shown by the arrows in the figure, when the amount of virus is low, a small amount of clear plaque is formed, just like bacteriophages. If the virus remains intact, all the host cells will die and there will be no more host cells. Therefore, the cells and plaque will not be stained with blue dye to make them easier to see.

Fig. 10-10 Glove box used in the spray test.

Fig. 10-9 Plaque of a virus infecting an animal cell.

In this glove box, a virus solution is sprayed to measure the antiviral effect of a small air purifier.

The area where the host cell is not dead is colored. Infection with a single viral particle causes the host cell to die. The host cells die and become transparent plaques (black arrows). The virus becomes visible. If the amount of virus is high, all host cells die (clear arrows). Translated with www. DeepL.com/Translator (free version).

4 Antiviral test method using glove box

For photocatalysis, no test method standard has been developed for antiviral activity in the atmosphere. Therefore, air purifiers equipped with photocatalysts are tested based on the standard prepared by the Japan Electrical Manufacturers' Association. Here, we introduce the performance evaluation of photocatalytic products using a glove box. The glove box should be sealed (Fig. 10-10) to prevent leakage of microorganisms to the outside, as they are sprayed. The test method involves placing the substrate to be tested in the prepared glove box and then spraying it with the target microorganisms using a nebulizer. After this, the test machine is put into operation and the number of microorganisms suspended in the atmosphere is checked for decrease. The method is confirmed by collecting the sprayed virus using a gelatin filter and measuring its infectious titer.

5 Other methods of evaluating antimicrobial performance

Other antimicrobial performance evaluation methods include those for molds and algae. All of these can be tested using the same method as for antimicrobial performance evaluation, but with different microbial species. However, there are differences in the cultivation and measurement methods of these microorganisms, so it is important to check the respective standards. In the case of algae, the amount of Chlorella is measured by absorbance, rather than cultured and measured. In the case of mold and algae, the intensity of UV light irradiation is higher than that of bacteria and bacteriophages, and the upper limit is raised to 0.8 mW/cm^2 for mold and 1.0 mW/cm^2 for algae. We have also increased the irradiation time to 24 hours. The reason for this is the structure of the microorganism. In the case of viruses, the destruction of the outer membrane and the loss of infectivity are the main reasons for their antiviral effect. Viruses are not able to reproduce themselves. For this reason, it is thought that a short period of time is sufficient to produce an antiviral effect if the appropriate photocatalyst is used, with a standard exposure time of 4 hours. Bacteria are capable of self-propagation and have a more complex structure than viruses, so the standard irradiation time for checking antimicrobial effect is 8 hours. Molds and algae, on the other hand, have a more robust self-propagating ability and structure and therefore require a longer irradiation time. Also, depending on the intensity of the UV light, each microorganism can withstand a different irradiation intensity and therefore a different temperature during the test. For these reasons, the appropriate irradiation intensity and duration are set for each microorganism.

6 Summary

As described above, there are methods to check the performance of photocatalysis various types of microorganisms. In this section, we introduced some points on performance evaluation for other microorganisms, focusing on the antimicrobial performance evaluation test method. At present, the global spread of the novel coronavirus is a major problem. Photocatalysis have been attracting a lot of attention for their safe and reliable antiviral effect, and domestic and international companies and research institutes are promoting research and development of antiviral products using photocatalysis. It has also been shown that photocatalysis have an effective antiviral effect versus the novel coronaviruses. Further research and development of new photocatalysts is expected in the future, and performance evaluation is becoming increasingly important. We hope that you will understand the outline of the antimicrobial test method in this chapter and refer to each standard for more detailed test methods to promote the research and development of photocatalysts.

Reference literature
1. "Microbes and Infectious Diseases that Changed the World" by Takeo Sakamaki (Shodensha, 2020)
2. Osamu Nakagome (supervisor), Shigeru Kamiya (editor), Tatsuo Tinya (editor), "Standard Microbiology, 13th Edition" (Igaku Shoin, 2018)

(Hitoshi Ishiguro)

Chapter **11**

Water splitting by photocatalytic activity

History of photocatalytic water splitting and latest research trends

Working principle of photocatalytic water splitting

Experimental methods of water splitting

Points that require special attention concerning the experimental results

Chapter 11-1

History of photocatalytic water splitting and latest research trends

Photocatalytic water splitting is expected to be a clean energy production method to produce hydrogen from water and light, and research is being conducted all over the world.

The origin of this research is the Honda-Fujishima effect, which was published in Nature in 1972. As shown in Fig. 11-1, an electrode system consisting of a rutile-type titanium dioxide single crystal and a platinum counter electrode was prepared, and when the titanium dioxide was irradiated with light, water was decomposed to produce hydrogen and oxygen. When titanium dioxide is irradiated with ultraviolet light, electrons in the valence band transit to the conduction band, and free holes and electrons are generated in the valence band and the conduction band, respectively. The holes in the valence band oxidize water and produce oxygen (O_2) and protons (H^+). On the other hand, the electrons in the conduction band move through an external circuit to the platinum counter electrode, where they reduce the protons to produce hydrogen (H_2).

In order to produce large quantities of hydrogen, it is necessary to prepare a large number of titanium dioxide electrodes. However, because single

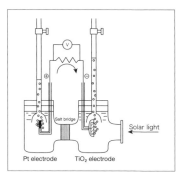

Fig. 11-1
Schematic diagram of the Honda-Fujishima effect.

crystals are expensive, a low-cost method of producing titanium dioxide electrodes had been demanded. Fujishima *et al.* exposed a titanium metal plate to a flame in air to form an oxide film on the surface, and used it as an electrode. The experimental system as shown in Fig. 11-2 was set up outdoors on a rooftop, and then seven liters of hydrogen were obtained per day from a one-meter square titanium dioxide plate. However, this amount is insufficient for practical applications. In addition, the solar-to-hydrogen conversion efficiency

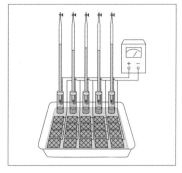

Fig. 11-2
Schematic diagram of the apparatus used in the experiment to produce hydrogen from water using sunlight.

Source: "All About Photocatalysis Revealed by Leading Experts", by Akira Fujishima (Diamond Inc., 2017).

(STHCE) was only 0.3%. This is because titanium dioxide can only use ultraviolet light, which is present at only a few percent of sunlight. In

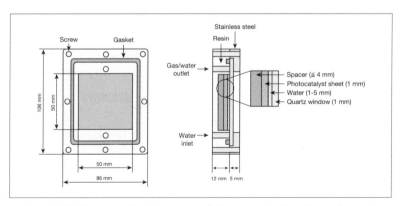

Fig. 11-3 Schematic diagram of the newly developed panel reactor.

Source: NEDO HP news release (January 19, 2018)
Available at: https://www.nedo.go.jp/news/press/AA5_100899.html

order to increase the STHCE, photocatalysts that can also use visible light, which is abundant in sunlight, are being actively developed, as explained in detail in the next Section.

For the practical application of hydrogen production by photocatalytic water splitting, it is important not only to increase the STHCE but also to decrease the size and cost of the reaction system. Water splitting experiments are usually conducted in a container with a certain depth, in which water is filled and photocatalysts powders are dispersed. However, if an enlargement of the reactor for practical use is considered, a large, strong reactor is required to hold a large volume of water, which increases the cost. Therefore, Domen *et al.* designed and developed a new photocatalytic panel reactor for water splitting, as shown in Fig. 11-3. The reactor contains a 50 mm square sheet of strontium titanate photocatalyst, which is supplied with water through a gap of only a few millimeters between it and a quartz window. Although strontium titanate photocatalyst, like titanium dioxide, is only available in ultraviolet light, it is capable of water splitting by itself. Therefore, the

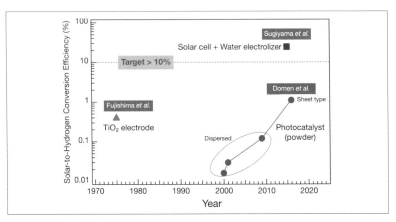

Fig. 11-4 Trends in solar-to-hydrogen conversion efficiency (STHCE).

panel on which strontium titanate is simply coated on the substrate functions as a photocatalytic panel. In experiments, water splitting was achieved even though a depth of water was only 1 mm, and the photocatalytic sheet remained active for more than 2 hours. In addition, tests with increased UV intensity demonstrated hydrogen generation corresponding to STHCE = 10%. Furthermore, they have succeeded in enlarging the scale to 1 m² and achieved STHCE = 0.4% in water splitting using natural sunlight.

Domen *et al.* have also developed a mixed powder photocatalytic sheet consisting of two types of visible light-responsible photocatalyst and conductive material fixed on a glass substrate, achieving STHCE = 1.1%. They have also developed a process for practical use and have succeeded in producing 10-cm square photocatalytic sheets using a screen-printing method that enables mass production.

Although studies on photocatalytic water splitting progress steadily, as shown in Fig. 11-4, the conversion efficiency is not sufficient yet. To realize the practical application of artificial photosynthesis, the STHCE should reach 10%, although the STHCE achieved in photocatalytic water splitting is only about 1.1%. For this reason, research on integrated systems combined with solar cells and photoelectrodes has been carried out vigorously. To give an example, Sugiyama *et al.* achieved STHCE = 24.4% in a hydrogen generation system combining a concentrator solar cell module and a water electrolyzer.

Working principle of photocatalytic water splitting

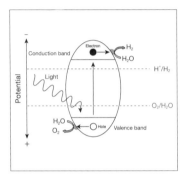

Fig. 11-5 Band structure of photocatalysts for water splitting.

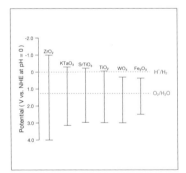

Fig. 11-6 Band structures of typical oxide semiconductor.

In the photocatalytic reaction, as a result of photoexcitation, electrons generated in the conduction band are used for reduction, and holes generated in the valence band are used for oxidation, and thus, in order to realize water splitting, as shown in Fig. 11-5, it is necessary that the potential of electrons in the conduction band is lower than the reduction potential of water (negative side), and the potential of holes in the valence band is higher than the oxidation potential of water (positive side). In the case of oxide semiconductors, such as titanium dioxide, the valence band is mainly formed from the $2p$ orbitals of oxygen. Therefore, as shown in Fig. 11-6, the position of the valence band of most oxide semiconductors is fixed at about 3.0 V relative to the Normal Hydrogen Electrode (NHE). On the other hand, since the reduction potential of water is 0 V, the conduction band must be more negative than 0 V. Therefore, a photocatalyst whose band gap is more than about 3 eV (about 400 nm in wavelength) is required, which would only allow the use of ultraviolet light, which is only a fraction of the solar spectrum. Since the ultimate goal of this research field is the practical application of solar hydrogen production from water using sunlight, it is desirable to develop a photocatalyst that could achieve water splitting under visible light irradiation.

One way to achieve visible light response is to narrow the band gap (Fig. 11-7). When an oxide semiconductor is doped with a small amount of a transition metal, a new impurity level is formed in the band gap (Fig. 11-7 (b)). By using optical excitation from this impurity level to the conduction band, water splitting can be achieved by irradiating with the longer wavelength light, which has a lower energy than the actual band gap.

In addition, by introducing metals (Cu^+, Ag^+, Pb^{2+}, Sn^{2+}, Bi^{3+}, etc.) that can form a new valence band at a shallower position than the valence band, consisting of the $2p$ orbitals of oxygen, or by replacing the $2p$ orbital of oxygen with the lower (negative) energy levels of the $2p$ orbital of nitrogen, the $3p$ orbital of sulfur or the $4p$ orbital of selenium, the valence band can be brought closer to the oxidation potential of water (Fig. 11-7 (c)). This has been used in some cases to achieve a visible light response, while maintaining a band structure that allows water splitting.

Another approach is to use a photocatalyst system known as a Z-scheme, which consists of a hydrogen evolution photocatalyst, an oxygen evolution photocatalyst and an electron transfer system that exchanges electrons between the two (Fig. 11-8). The structure of the photocatalyst mimics the photosynthetic mechanism of plants. The band gaps of the hydrogen and oxygen evolution photocatalysts are small, allowing water splitting using longer wavelengths of light.

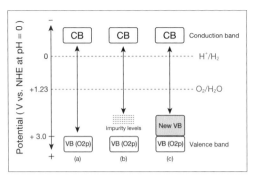

Fig. 11-7 Conceptual diagram of band gap narrowing.

(a) Original band structure, (b) Introduction of impurity levels, (c) Creation of shallow valence band

Nowadays, photocatalysts are being developed that are capable of water splitting even with red light (about 700 nm wavelength).

For the practical application of solar hydrogen production, it is necessary to increase the STHCE. To achieve this, not only extending the wavelength range of light that can be used for the water splitting reaction, but also ensuring that the absorbed light can be used efficiently for the reaction is important. To achieve this, optimization of the synthesis process of the photocatalyst and co-catalyst (metal or metal oxide particles supported on the photocatalyst as active sites in the redox reaction) are important. To give an example, Domen *et al.* controlled the particle morphology of strontium titanate photocatalysts and selectively introduced hydrogen-generation and oxygen-generation co-catalysts at specific crystal facets. By selectively transferring the electrons and holes generated by photoexcitation to the respective co-catalysts, recombination is suppressed, and almost

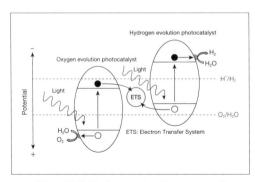

Fig. 11-8 Schematic diagram of Z-scheme photocatalyst system.

all of the absorbed light can be used for water splitting reactions.

Chapter 11-3

Experimental methods for water splitting

In order to accurately evaluate the amount of hydrogen and oxygen produced by photocatalytic water splitting, it is necessary to evaluate them in a closed circulation system, as shown in Fig. 11-9. The system consists of a reactor, a vacuum line and a gas sampling port directly connected to the gas chromatograph. As discussed below, the quantitative detection of oxygen gas is very important when evaluating photocatalytic water splitting; therefore, it is important to ensure that no air is introduced into the measurement system.

Although reaction vessels in a variety of shapes and sizes can be considered, an internal illumination-type condenser is ideal for efficient light irradiation. If the photocatalyst has a large band gap and needs to be irradiated at wavelengths below 300 nm, high-pressure mercury

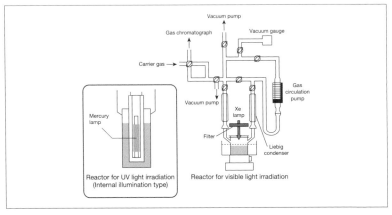

Fig. 11-9 Schematic diagram of a closed environmental system.

lamps are generally used as the light source. If visible light irradiation is desired, xenon lamps and UV-cut-off filters are used. If hydrogen production under sunlight is considered, a solar simulator is the standard light source.

Points that require special attention concerning the experimental results

When evaluating the results of experiments on the complete photocatalytic decomposition of water, there are several points to pay attention.

First of all, it is necessary to confirm that hydrogen and oxygen are produced in the stoichiometric ratio (hydrogen: oxygen = 2:1). If only hydrogen production occurs, the amount of hydrogen produced is often very small in relation to the amount of photocatalyst, and it is difficult to conclude that it is photocatalytic hydrogen production. On the other hand, since the complete decomposition of water is very difficult, reactions with sacrificial reagents are often used to evaluate photocatalytic activity. When a reducing agent such as alcohol is added, the holes in the valence band are consumed and only hydrogen is produced as a result of the reduction of water. In contrast, when an oxidizing agent such as silver chloride is added, the electrons in the conduction band are consumed and only oxygen is produced as a result of the oxidation of water (see Fig. 11-5 for the relationship between the electrons and holes produced in the photocatalyst and the production of hydrogen and oxygen).

Secondly, it is necessary to check that the production of hydrogen and oxygen increases with time of light irradiation. If the rate of production decreases with time, the reaction may be due to impurities.

It is also important to check that the total amount of product (hydrogen and oxygen) exceeds the amount of catalyst. If the total product is less than the amount of photocatalyst, other non-photocatalytic stoichiometric reactions are suspected to have occurred. This can be assessed using the turnover number (TON), which is defined as follows:

TON = (number of reacted molecules) / (number of active sites)

However, as it is difficult to determine the number of active sites, the following formula is used:

TON = (number of reacted electrons) / (number of atoms in the photocatalyst)

or

TON = (number of reacted electrons) / (number of atoms on the surface of the photocatalyst)

The number of reacted electrons is calculated from the amount of hydrogen gas produced. Photocatalytic water splitting has occurred if the TON is above 1, while other reactions should be suspected if the TON is much below 1. In general, photocatalytic activity is not proportional to the mass of the photocatalyst used and should not be converted to activity per gram (e.g., μ mol h^{-1}g^{-1}).

Since the activity of a photocatalytic reaction depends on the experimental conditions, such as light intensity and the type of the reactor, it is not possible to compare the reaction efficiency between experiments with different experimental conditions. Therefore, the quantum yield (QY), defined by the following equation:

QY (%) = ((number of reacted electrons) / (number of absorbed photons)) × 100

is important for the comparison between experiments. However, since it is difficult to measure the number of photons absorbed by the photocatalyst, the apparent quantum yield (AQY), which is replaced by the number of incident photons, calculated as follows:

AQY (%) = ((number of reacted electrons) / (number of

incident photons)) × 100

is used. If this value is extremely small, the tested material should be examined carefully to see whether it can truly be considered to be a photocatalyst. Note that the quantum efficiency is different from STHCE, which is defined by the following equation:

$$STHCE(\%) = \left(\frac{\text{Energy output as hydrogen}}{\text{Energy of incident sunlight}} \right) \times 100$$

$$= \left(\frac{r_{H2} \times \Delta G}{P_{sun} \times S} \right) \times 100$$

Energy output as hydrogen is calculated from the rate of hydrogen production r_{H2} (mmol s^{-1}) and the gain in Gibbs energy ΔG (237 kJ mol^{-1}), while the energy flux of the sunlight P_{sun} (W m^{-2}) and the area of the reactor S (m^2) is used to calculate the energy of incident sunlight. If AM 1.5 solar spectrum is considered, its energy flux (P_{sun}) is 1 kW m^{-2}. If hydrogen production using sunlight is the target, STHCE evaluation is required.

Finally, it is necessary to determine that the water splitting reaction proceeds when light of the proper wavelength (higher energy than the band gap) is irradiated and stops in the dark (photoresponsivity). This requires a comparison of the absorption spectrum of the photocatalyst with the wavelength dependence of the photocatalytic reactivity (action spectrum). To measure the action spectrum, it is necessary to irradiate monochromatic light (light of a specific wavelength) using a bandpass filter or interference filter. This data is particularly essential when discussing visible light-responsive photocatalysts, because absorption in the visible light region sometimes cannot contribute to the photocatalytic reaction. In addition, some metal oxides undergo water splitting by mechanocatalysis (catalytic reaction using mechanical reaction) simply by stirring in water without light irradiation; therefore, it is necessary to

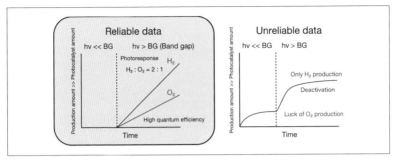

Fig. 11-10 Points on the experimental data.

conduct control experiments in the dark to confirm that the water splitting reaction is truly photocatalytic.

Fig. 11-10 shows examples of reliable experimental data and data that require caution.

References

1. "Heterogeneous photocatalyst materials for water splitting" Akihiko Kudo, Yugo Miseki, Chemical Society Reviews, 38, 38, 253-278 (2009)

(Norihiro Suzuki)

Chapter 12

Future Prospects

Water purification
Agricultural applications
Medical applications
Protecting historical landmarks and handcrafted items
Indoor applications
Research trends in artificial photosynthesis

Chapter 12-1

Water purification

The ideal use of photocatalysts in environmental purification is to remove trace pollutants using solar energy. The density of solar energy on the earth's surface is 1 kW/m2, and pollutants are typically present in a dilute form in the environment. Both China and South Korea are actively engaged in national air purification initiatives, and in Japan, a photocatalyst-containing pavement was developed in the late 1990s to remove nitrogen oxides from road surfaces using the "photo-road" method. In this way, the contribution of photocatalysis to environmental purification is actively investigated in outdoor areas where the use of solar energy can be maximized.

In 2020, coronavirus was reportedly detected in sewage, underscoring the need to ensure the safety and security of environmental water, including untreated effluents. The involvement of photocatalysis is also expected in this area. However, unlike air, water is difficult to collect in

Fig. 12-1 Example of large-scale water purification using a photocatalytic net.

(Source: http://www.slgpt.com/)

the processing facility and treat simultaneously because of its resistance, and the efficacy of photocatalysis is impaired by light blockage. Moreover, in lakes, ponds, and marshes, where water flows slowly, waterborne pollutants cannot be successfully absorbed by photocatalysts, necessitating a different approach from that used for removing pollutants from air.

Further progress in the research and development of photocatalysts is required for treating large volumes of water. A Chinese company, Shuangliang, has developed a lightweight, unpowered photocatalytic net that can float on the surface of lakes and ponds to purify water on a large scale (Fig. 12-1). Although the efficacy, durability, affordability, and environmental impact of this photocatalytic net require evaluation, this device can be expected to contribute to water purification on a global scale.

Solutions for small-scale water treatment are outlined in Chapter 9. In recent years, development of ultraviolet light-emitting diodes has progressed, and with their long life and compactness as light sources, photocatalytic technology can be incorporated into water purification systems on a small scale, creating various product categories.

Agricultural applications

Food loss has become a social problem in developed countries such as Japan, and food shortage is projected to become a major concern worldwide. These seemingly contradictory issues can be solved by digitally managed smart agriculture, which introduces a vast business opportunity. Smart agriculture incorporates information and communication technologies, robotics, and artificial intelligence, and efficiency in smart agriculture requires controlled spaces such as greenhouses, where temperature and humidity can be maintained at optimum levels, unlike in open field cultivation. However, even transparent surfaces can become dirty, reducing penetration by sunlight, which is necessary for plant growth. Photocatalytic coatings are used on the exterior-facing side of building materials to block heat from sunlight and maintain residences at a comfortable temperature. These photocatalytic coatings could also be used on the transparent plastic of greenhouses to dissipate heat while allowing in the light necessary for photosynthesis. Achieving these effects requires the use of a material with excellent photocatalytic properties that does not damage polymeric vinyl. The sunlight that reaches greenhouse plants can also be reduced by condensation forming on the inside of greenhouse walls, and when water condenses on foliage, plant material may rot. Photocatalysts with superhydrophilic properties can prevent condensation, thus contributing to plant growth (Fig. 12-2a).

Glasshouses, which are more durable than plastic greenhouses and can be located outdoors, are constructed entirely of glass that can be self-cleaned with photocatalysts, unlike polymeric vinyl. By maintaining a

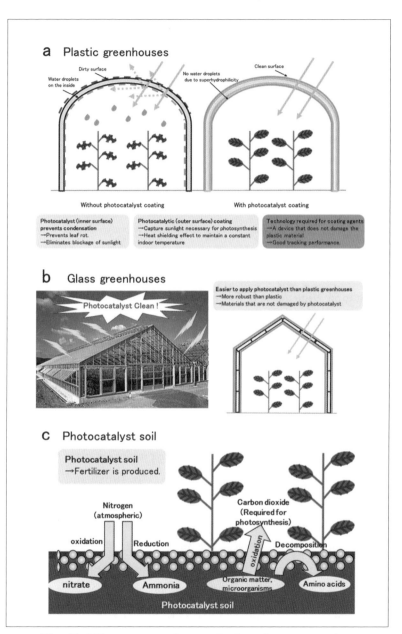

a Plastic greenhouses

Dirty surface

Clean surface

Water droplets on the inside

No water droplets due to superhydrophilicity

Without photocatalyst coating

With photocatalyst coating

Photocatalyst (inner surface) prevents condensation
→Prevents leaf rot.
→Eliminates blockage of sunlight

Photocatalytic (outer surface) coating
→Capture sunlight necessary for photosynthesis
→Heat shielding effect to maintain a constant indoor temperature

Technology required for coating agents
→A device that does not damage the plastic material
→Good tracking performance.

b Glass greenhouses

Photocatalyst Clean!

Easier to apply photocatalyst than plastic greenhouses
→More robust than plastic
→Materials that are not damaged by photocatalyst

c Photocatalyst soil

Photocatalyst soil
→Fertilizer is produced.

Nitrogen (atmospheric)

Carbon dioxide (Required for photosynthesis)

oxidation   Reduction   oxidation   Decomposition

nitrate   Ammonia   Organic matter, microorganisms   Amino acids

Photocatalyst soil

Fig. 12-2 Next-generation agricultural system using photocatalysis.

clean glass surface and constant transmission of visible, near-infrared, and far-infrared light, it is possible to create an indoor environment with excellent moisture retention (Fig. 12-2b). In glasshouses, photocatalytic technology can be applied at an earlier stage and is expected to be widely used.

Other technologies nearing the stage of practical use include photocatalytic treatment of pesticide effluents and hydroponic nutrient solution treatment. These can be used in small water treatment systems or applied in devices in which direct sunlight is irradiated to greenhouses. Experiments carried out by Yoko Miyama and colleagues at the Kanagawa Agricultural Technology Center have shown that photocatalytic water treatment systems are effective in purifying agricultural effluents, and like the self-cleaning glasshouses, this is a technology that is in the process of being actively industrialized.

Plant factories have also attracted attention for their capacity to produce crops in a highly controlled environment. The most common type is hydroponics, where crops are grown in nutrient-rich water without the use of soil. Unlike soil-based cultivation, weed control is required less frequently in hydroponics and plants can be nourished directly to increase the yield. However, one problem in plant factories is contamination of the growth solution by fungi and algae. In hydroponics, the nutrient solution is circulated through the system, and contamination becomes apparent during the cultivation process. Algal blooms on the growing racks can impair plant growth as the algae absorb the nutrients, and some algae are toxic to plants, inhibiting growth and causing disease; this results in reduced yield and quality. Photocatalysts could therefore offer a method for purifying the growth medium and improving production efficiency.

Nitrogen fixation by photocatalysis is also attracting attention as a promising technology, albeit at a basic research level. This technology uses photocatalytic reactions to convert nitrogen, which makes up 78% of the atmosphere, into ammonia and other compounds using sunlight. Ammonia can be synthesized using the reduction reaction of the photocatalysis, and nitrate can be produced using the oxidation reaction and then used as a fertilizer. Photocatalysts manufactured in bulk and sprinkled on soil could react with sunlight to produce fertilizer using atmospheric nitrogen as a raw material. The photocatalyst, which is inorganic, would not be consumed in the reaction and could continue to produce fertilizer. Furthermore, photocatalysts applied to the soil could break down organic matter and microorganisms to yield fertilizers such as amino acids. Alternatively, the complete decomposition of the organic matter might lead to the release of carbon dioxide and the enhancement of photosynthetic reactions. As research and development of photocatalytic materials progress, so too would the applications. Other smart agriculture practices using photocatalysts could include filtering culture media for plant growth (Fig. 12-2 c).

Medical applications

Photocatalysis are expected to be used in healthcare settings to prevent infection in applications such as photocatalytic tiles in the walls of operating rooms. Other medical applications that leverage the superhydrophilic properties of catalysts include anti-fogging lenses for endoscope cameras and catheters and syringes with a low coefficient of friction. Photocatalysis are also being studied in the sterilization of catheters. Photocatalysis have been shown to kill bacteria, viruses, and the HeLa cervical cancer cells (Fig. 12-3), which may indicate their potential as antineoplastic agents.

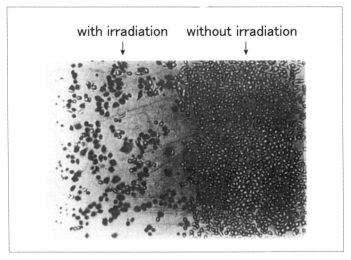

Fig. 12-3 Photocatalytic killing effect on HeLa cervical cancer cells.

Photocatalytic nanoparticles have been reported that are active against radiation. Unlike conventional photocatalysts that respond to light, photocatalysts that can produce reactive oxygen species when exposed to X-rays could exert a therapeutic effect in internal organs by specifically recognizing neoplastic cells and irradiating them (Fig. 12-4).

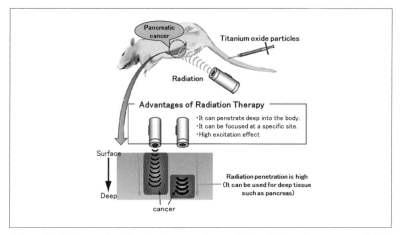

Fig. 12-4 Hypothesized effect of combined titanium peroxide/ radiation therapy.

Hand, foot, and mouth disease is caused by enteroviruses and is prevalent in the summer months, especially in infants. The main symptom is a rash that appears in the mouth and on the hands and feet (Fig. 12-5) and in rare cases can lead to complications such as meningitis. The infection is transmitted by droplets from sneezing and through the feces and is therefore likely to occur at daycare facilities. It is possible that photocatalytic coatings of toys, picture books, towels, and other items in common use around infants and young children may prevent transmission; the application may also be extended to the prevention and control of other infectious diseases.

The use of photocatalysts is also being researched in the field of dentistry. Whitening teeth by titanium dioxide photocatalysts and blue light-emitting diodes is already in practical use (Fig. 12-6), and the

Fig. 12-5 Location of rash in hand, foot, and mouth disease.

technology may be extended in the future to dental implants (artificial tooth roots), dentures, and denture-cleaning agents, as well as to infection prevention through oral care. Water jets are used in the process of tooth extraction and to clear the patient's mouth; photocatalytic additives would ensure the cleanliness of the water in an easier and less costly manner than hypochlorite or ozonation.

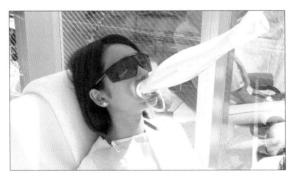

Fig. 12-6 Example of tooth whitening (Charion Co., Ltd.).

Chapter 12-4

Protecting historical landmarks and handcrafted items

Nikko Toshogu is a wooden structure dating from Japan's Edo period (1603–1868), and is a World Heritage Site decorated with lacquer and intricate, colorful carvings. However, lacquer deteriorates and fades when exposed to ultraviolet light, and the glue used to mix the pigment particles is susceptible to mold (Fig. 12-7). Titanium dioxide photocatalysts can absorb ultraviolet light, and such a coating, if successfully applied to the surface of lacquered wood, would prevent fading and mold growth. Lacquer finishes have a unique and elegant aesthetic quality and are a traditional Japanese craft for producing fine tableware, furniture, and musical instruments. A transparent coating of photocatalysts on lacquer would extend the life of lacquered items, especially those located outdoors. The self-cleaning function of photocatalysis maintains the aesthetics of the item coated and prevents bacterial and fungal damage.

Fig. 12-7 Moldy decoration at the Nikko Toshogu shrine, Japan.

Indoor applications

Development of photocatalysis that are highly sensitive to visible light has become popular. Quantum-mechanical calculations have shown that if some oxygen atoms in titanium dioxide crystals are replaced by other elements that, like oxygen, tend to become anions (negatively charged ions), the crystals respond to visible light. Such elements include nitrogen, sulfur, and carbon, and they have been used to develop photocatalytic materials that respond to visible light. Other crystals that have been tested include tantalum oxynitride, tungsten oxide, and composites of titanium dioxide and tungsten oxide.

Tungsten oxide has been known for some time to exert photocatalytic activity, but this has not been put to practical use because of insufficient sensitivity. Recently, however, Toshiba and other companies have achieved high sensitivity of tungsten oxide to visible light by adjusting the crystal structure to increase the efficiency of charge separation and reduce particle size to increase the contact area with the target material, leading to significant progress in the application of tungsten oxide.

Figure 12-8 shows the range of visible light absorbed by tungsten oxide and the spectral distribution of white light from light-emitting diodes. Tungsten oxide can be expected to be sensitive to white light-emitting diodes, which are becoming increasingly popular, and is being introduced into various materials by kneading or coating onto fibers, for applications in hotels, hospitals, nursing homes, daycare facilities, and automobile interiors.

Photocatalysis are also expected to be used as purification agents on spacecraft and space stations, where unpleasant odors may linger in the air. Photocatalysis could be used to deodorize and sterilize the interior of spacecraft and space habitats.

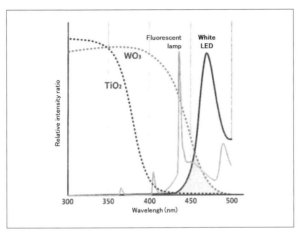

Fig. 12-8 Range of visible light that can be absorbed by tungsten oxide (WO₃) and spectral distribution of white light from light-emitting diodes (LEDs).

Source: "All About Photocatalysis Revealed by Leading Experts" by Akira Fujishima (Diamond Inc., 2017)

Research trends in artificial photosynthesis

Research on artificial photosynthesis using photocatalysts to produce solar fuels such as hydrogen from water or solar chemicals such as methanol from carbon dioxide is active and ongoing (Chapter 11). Artificial photosynthesis has been actively studied for half a century, since the discovery of the Honda-Fujishima effect, and recent environmental and energy problems further promote this area of investigation. Figure 12-9 shows the number of citations of the 1972 paper by Fujishima and Honda, which began to increase gradually in the first half of 2000. The number of citations of the *Nature* paper was so high that Akira Fujishima was named a Thomson Reuters Citation Laureate in 2012.

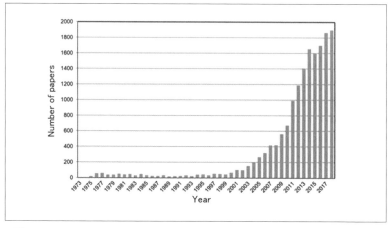

Fig. 12-9 Number of citations for A. Fujishima and K. Honda, Nature, 238 (1972) 37.

Fujishima and Honda showed that light energy can be used to decompose water to oxygen and hydrogen. However, the fact that titanium dioxide can only absorb near-ultraviolet light limits its capacity to produce hydrogen efficiently using sunlight and has triggered a worldwide search for photocatalytic systems that can harness the visible light from the sun by modifying titanium dioxide or using other metal oxides or metal nitrides. For example, artificial leaves developed by Daniel Nocera's group at Harvard University can decompose water and extract hydrogen using sunlight. These leaves consist of a silicon solar cell with a catalyst on both sides. One side of the semiconductor silicon is coated with a catalyst made of cobalt and other materials, and the other side contains an alloy catalyst of nickel, molybdenum, and zinc. Practical application of this technology is currently limited by costs, durability, and efficiency.

Materials containing metal oxides are considered advantageous for practical use because they can be synthesized relatively easily and in large quantities and can therefore be applied over large areas at low cost. Active research is being conducted using both semiconductors and metal complexes or composites containing metal complexes. However, metal oxides exhibit insufficient absorption of sunlight, limiting their efficiency, but one material that responds to light up to about 500 nm and can completely decompose water has been reported. One of the example is metal oxynitrides, which is one type of photocatalysts and works as a single-particulate photocatalyst for water splitting under visible light irradiation. Materials that combine two types of photocatalysts are termed Z-scheme photocatalysts (Fig. 11-8) and consist of a hydrogen-producing photocatalyst particle that can absorb low-energy light, an oxygen-producing photocatalyst particle, and an electron-transfer system that exchanges electrons between the two particles. The flow of photo-excited electrons resembles the letter Z, hence the name. This mechanism enables water splitting using visible light at long wavelengths.

With a view to the practical application of the Z-scheme photocatalyst, an attempt has been made to immobilize it in sheet form (Fig. 12-10). This photocatalyst has a simple structure that is suitable for application over a large area and at low cost and has the potential to supply inexpensive hydrogen on a large scale. The findings of research into this photocatalyst fixed in sheet form have been reported by the Research Association for Artificial Photosynthetic Chemical Process Technology, which was organized by a number of major Japanese manufacturers and research institutes. Large-scale research and development efforts are continuing with government support.

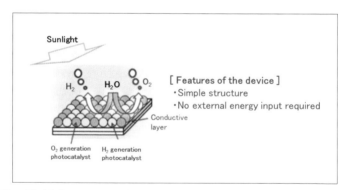

Fig. 12-10 Conceptual diagram of water splitting by mixed-powder photocatalyst sheet.

The New Energy and Industrial Technology Development Organization is currently developing a basic technology for producing clean hydrogen from water by photocatalysis using solar energy, in combination with the development of separation membranes and synthetic catalysts and the production of key materials such as olefins from carbon dioxide. For practical application of artificial photosynthesis, the goal is to achieve a solar hydrogen conversion efficiency of 10%. Figure 12-10 shows results of approximately 1.1% in 2016, but recent work has achieved higher conversion efficiency. Hybrid systems that combine solar cells and water electrolysis cells have been reported in many cases to exceed a conversion efficiency of 10%, and some to exceed 20%. Challenges to practical application include efficiency, cost, and durability.

Another trend is accelerated research on carbon recycling, in which carbon dioxide generated in thermal power plants is reduced by photocatalysts and converted into useful chemical materials and fuels (Fig. 12-11). The photosynthetic efficiency of plants is difficult to quantify but is estimated to be around 0.2% for a typical plant. Thus, the direct conversion efficiency of carbon dioxide is more challenging than that of hydrogen, but artificial photosynthesis is clearly the key to

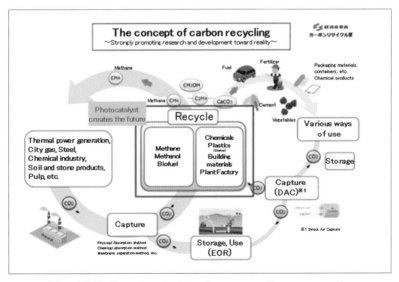

Fig. 12-11 Accelerated carbon recycling research.

solving resource, energy, and environmental problems and will involve the participation of the next generation of researchers and a wide variety of technologies. Moreover, research focus should extend from immediate efficiency to a long-term and comprehensive picture.

(Chiaki Terashima)

Chapter

13

Photocatalyst Museum

Photocatalyst Museum

The Kanagawa Institute of Industrial Science and Technology (KISTEC, formerly Kanagawa Academy of Science and Technology) opened the Photocatalyst Museum in July 2004, the only facility in Japan to exhibit photocatalytic technology, in the hope that people would learn about photocatalysis as a familiar technology.

In the Photocatalyst Museum, there is a demonstration device that reproduces the Honda-Fujishima effect, the principle of photocatalysis. When light (ultraviolet light) hits the surface of a titanium dioxide electrode, oxygen is produced from the surface, and hydrogen from the platinum electrode (the counter electrode).

There are also panels on display explaining the principles of photocatalysis in an easy-to-understand way. If you ask when you enter

Entrance to the Photocatalyst Museum.

The "Honda-Fujishima Effect" device.

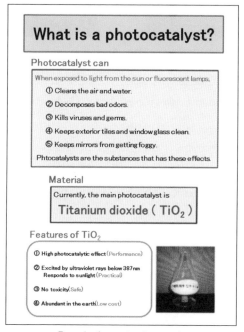

What is a photocatalyst?

Photocatalyst can

When exposed to light from the sun or fluorescent lamps,

① Clears the air and water.

② Decomposes bad odors.

③ Kills viruses and germs.

④ Keeps exterior tiles and window glass clean.

⑤ Keeps mirrors from getting foggy.

Phtocatalysts are the substances that has these effects.

Material

Currently, the main photocatalyst is

Titanium dioxide (TiO_2)

Features of TiO_2

① High photocatalytic effect (Performance)

② Excited by ultraviolet rays below 387nm Responds to sunlight (Practical)

③ No toxicity (Safe)

④ Abundant in the earth (Low cost)

Example of an explanatory panel.

the museum, the staff at the reception desk will be happy to show you around and give you an explanation at a time to suit you.

In addition, by placing actual photocatalytic products on display, the museum introduces the market situation and applications of

Tent with photocatalyst

Interior and exterior materials (tiles, etc.)

Experimental equipment for removing stains from photocatalytic exterior walls

Director Fujishima actually applying the coating agent

photocatalytic products, which have been developed by manufacturers, trading companies and business organizations that deal with photocatalytic products. Examples of products on display include coating agents, interior and exterior materials (tiles, etc.), outdoor products (tents), air purification filters and air purifiers.

For business visitors who wish to gather information on the use of photocatalytic technology in their own products, or to find a partner who can provide suitable materials, the museum displays relevant

Group tour of third-grade elementary school students

Picture book corner

| Brochures | Business meeting space |

brochures in an easily accessible location and also makes arrangements with partner organizations. They can also bring their business partners to visit the museum and use the business meeting space in the museum free of charge.

The Photocatalyst Museum also aims to promote interest in photocatalytic technology and science in general among the local community by organizing experiments for children, training for local science teachers and group visits. At the entrance to the Photocatalyst Museum, there is a picture book corner where visitors can browse through 600 picture books and children's books donated by Dr. Fujishima, as well as related organizations, in addition to specialist books for adults. We aim to provide a place where people of all ages can come into contact with science through familiar objects, just as if they were reading a picture book.

Since its opening, the museum has welcomed more than 110,000 visitors (as of December 2020), ranging from young children to adults and experts. The annual number of visitors is between 4,000 and 5,000, and the museum has also received group visits from companies in the chemical, electrical equipment, housing and construction materials, steelwork and other manufacturing industries; overseas visitors from the

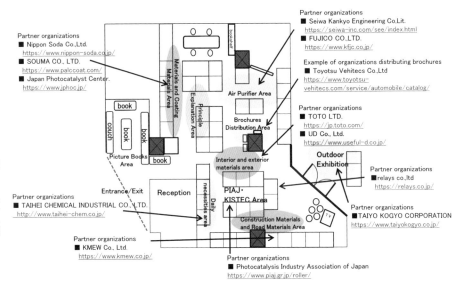

USA, China, Korea, India, Germany and Australia; students from universities, technical high schools, junior high schools and primary schools; and government officials and citizens' groups. We would like to express our gratitude to all visitors.

The museum will continue to contribute to the development of the photocatalysis industry and to provide a place where many people can enjoy science.

● Photocatalyst Museum

3-2-1 Sakado, Takatsu-ku, Kawasaki City, Kanagawa Prefecture 213-0012, Japan
1st Floor, West Wing, Kanagawa Science Park (KSP)
Kanagawa Institute of Industrial Science and Technology (KISTEC)

Please note that due to the measures against the novel coronavirus infection, the number of group visitors, the list of books and some demonstration experiments are restricted. We will change the services available depending on the situation in the future.

(Tomoko Aoki and Momoko Tsurumi)

Chapter **14**

Dissemination of Photocatalysis

The Photocatalysis Industry Association of Japan
List of members
Kagoshima photocatalysis construction association

The Photocatalysis Industry Association of Japan

1. What is the Photocatalysis Industry Association (PIAJ) of Japan?

Titanium oxide photocatalysis has attracted attention for its various applications such as antifouling, anti-fogging, antibacterial, antifungal and air and water purification, so active research and development is being carried out in universities, various research institutes and private enterprises. Moreover, as products in which this technology has been applied have begun to be put into practical use in different fields of daily life, construction and civil engineering, considerable expansion of this market is predicted, or rather expected, in the future.

On the other hand, however, further improvement and standardization in quality and performance of photocatalyst materials and their application for products is required. In consideration of this situation, the PIAJ was established to:

- call for participation of various private enterprises in producing and selling photocatalyst materials and products,
- grasp problems of the market reflecting the opinions and needs of users, with the approbation and participation of related government ministries and agencies, as well as of corporate users of photocatalyst products in the fields of daily life, construction and civil engineering,
- perform a variety of activities to solve the above problems by

working with the Photo Functionalized Materials Society, and

- foster a sound market for photocatalyst products and promote them.

2. Purpose of establishment

Our purpose is to contribute to the enhancement of people's standard of living and the growth of related industries, by promoting the diffusion of these products through application, expansion, and public relations activities related to photocatalytic technology, and by fostering the creation of a sound market through technological improvements and supply of high-quality products.

3. Activities

- Promoting the standardization of photocatalytic products.
- Promoting the improvement of quality, performance and safety of photocatalytic products and their performance indication.
- Promoting market acknowledgement of photocatalytic products.
- Promoting the increased application and diffusion of photocatalytic technology.
- Exchange and cooperation with agencies and corporations related to photocatalysis.
- Cooperation with consumer organizations.
- Investigation, research, public relations, and the organization of lectures and workshops concerning items 1 to 4 listed above.
- In addition to the preceding articles, any activity which may be necessary for the attainment of our purpose.

4. Logo

The Photocatalysis Industry Association of Japan's logo is an image of water with ripples changing from blue to green, with an eye-catching yellow-orange accent as an image of a sunbeam. This is representative of photocatalysis and its capability forb environmental purification.

Fig. 14-1 Photocatalysis Industry Association logo.

This logo is applied to documents published by the industry association, and also used to show that an enterprise is an association member.

However, this logo does not prove that an individual product is authorized by the industry association. We have established standards for logo usage, and have banned misleading usage for general consumers.

5. PIAJ mark

The PIAJ mark (Fig. 14-2) is a certification mark that the Photocatalysis Industry Association of Japan gives to photocatalyst products whose performance, usage and other features are recognized as appropriate by the association. The association adopted JIS test methods as the yardstick to measure photocatalysis performance and set specific

Fig. 14-2 PIAJ mark.

performance standards through verification and examination from diverse angles. In addition, the association obtained feedback about performance standards from customers and governmental agencies, and established the standards based on the feedback. The PIAJ mark is given to photocatalyst products that meet the performance standards established in this way. Although the mark is currently given to products that are manufactured in Japan, the association plans to make the mark available in Asia and throughout the world in the future.

6. Performance criteria of certification

In product certification, in which the Photocatalysis Industry Association (PIAJ) grants the PIAJ mark to photocatalytic products that are recognized as appropriate in terms of performance and usage, the subject products must satisfy certain performance criteria in the performance evaluation using JIS test methods.

The performance criteria of certification and test methods for each function of photocatalysis for which the Photocatalysis Industry Association certifies products can be viewed on the Photocatalysis Industry Association's website.

7. The Photocatalysis Industry Association of Japan website

The website of the Photocatalysis Industry Association of Japan (PIAJ) has a wide range of information from introductory information, such as the principles of photocatalysis and a glossary, to information for developers, such as performance evaluation methods and recommended testing institutes, in addition to the purpose of establishment and activities of the association as explained in this document. In addition, the PIAJ has a unique feature that allows users to search and view information on the 107 active members (as of 7 January 2021, see Appendix of this chapter) and their registered products. In addition to that, you can search for products that have obtained the PIAJ mark by keyword or function and browse information on their performance and test methods. In this way, the website of the Photocatalysis Industry Association of Japan is constantly updated with useful information for all users of photocatalysts, from consumers to developers. Information on recommended institutions for products and tests is also available in the Appendices of this publication in Chapters 3, 6 and 8.

The Photocatalysis Industry Association of Japan
https://www.piaj.gr.jp/en/

List of members

(A)
Aderans Co.Ltd
AIR WATER INC.
ariel corporation
Art Creation Co., Ltd.
Asahi Kasei Corporation
Asakawa Kankyo Giken CO.,
    Ltd.
AsukaTech Co., Ltd.
A. G. T Co., LTD

(B)
B&B Co., Ltd.

(C)
CATARISE Corp.
Chemical Tecnology Co., LTD,
Chukoh Chemical Industries,
    Ltd.
Chuo Kankyo Sousetsu Co., Ltd.
COLOR Inc.
Credo Japan

(D)
Dai Nippon Printing Co., Ltd.
DAIKO ELECTRIC CO., LTD.
Danto Corporation

(E)
Ecoat Co., Ltd.

(G)
Gaea Co., Ltd.
GEN GEN CORPORATION

GODAI CHEMICALS INC.
GOOD HOME Co., Ltd

(H)
HIRAOKA & CO., LTD.
HIROSE MATAICHI CO.,
    LTD.
HOT FIELD Co., Ltd.
H · M ENGINEERS Co., Ltd.

(I)
Iris Corporation
Ishihara Sangyo Kaisha, Ltd.

(J)
Japan Photocatalyst Center
Japan Art
jp-corporatin Co., LTD

(K)
KIYOHARA & Co., Ltd.
Kawamori Co., Ltd.
KF Chemicals, Ltd.
KMEW Co., Ltd.
Koyo Engineering Co., Ltd.
KOYU Co., Ltd.

(L)
LIXIL Corporation

(M)
MAJAPAN Co., Ltd
MARUSYO SANGYO CO.,
    LTD.

Marutomi Y.K.
MIHASI.CO.,LTD.

(N)
Nagamune Corporation Co., Ltd
NANOBEST JAPAN Co., Ltd.
NANOWAVE Co., Ltd.
NEC Fielding, Ltd.
NIKKO INDUSTRY
    CORPORATION
NIPPON AEROSIL CO., LTD.
NIPPON NANOTEC Inc.
Nippon Paint Holdings Co., Ltd.
Nippon Soda Co., Ltd.
NP Corporation

(O)
Okitsumo Incorporated
Opeth

(P)
Pal Messe Co. Ltd.
Panasonic Corporation
PGS HOME Co., Ltd.
Photo-Catalytic Materials Inc.
Pialex Technologies Corp.
Polymer Holdings Co., Ltd.

(R)
Renatech Co., Ltd.
Ruden Bldg. Management Co.,
    LTD.

(S)
KABUSIKIKAISYA SHIRAISHI
s-grow Co., Ltd.

Seiwa Inc.
Sekistone Co., Ltd.
SEKISUI HOUSE, LTD.
Seven Chemical Co., Ltd.
Shin-Etsu Astech Co., Ltd.
Shin-Etsu Chemical Co., Ltd.
Showa Ceramics Co., Ltd.
SOUMA CO., LTD.
SUM
SunBless CO., LTD.
SUNTYPE CO., LTD.

(T)
Taiyo Kogyo Corporation
Takahara Group
    http://www.takahara-corp.jp/
Taki Chemical Co., Ltd.
Tayca Corporation
TOHO SHEET&FRAME Co.,
    Ltd.
TOPRUN
TOTO Ltd.
TOTO OKITSUMO Coatings
    Co., Ltd.
Toyokosho Co., Ltd.
Toyota Tsusho Corporation
Tsuruya Co., Ltd.
TWO Co., Ltd

(U)
UD Co., Ltd.
(W)
Wako Filter Technology Ltd.

(Y)
YKK AP Inc.

Kagoshima photocatalysis construction association

There is a local association that is actively involved in construction using photocatalysts, the Kagoshima photocatalysis construction association, which has been in operation for more than 10 years. Here is the outline of the association.

The association was established in September 2009 with the aim of contributing to safety and security by applying photocatalysts, which have excellent antibacterial and antiviral effects, to living spaces and to spread the use of photocatalysts. They use photocatalytic coating materials developed by companies in Kagoshima Prefecture (e.g., Eden Paint, see p. 129), with the cooperation of professors from Kagoshima University.

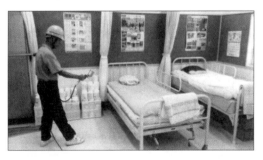

In October 2018, the 10th anniversary of the foundation was celebrated with a lecture by the author Akira Fujishima. The event was attended by

Fig. 14-3 Example of implementation in Kagoshima prefecture.

150 people, including the governor and members of the prefectural assembly.

An example of construction work is shown in Fig. 14-3.

Currently, 29 companies in Kagoshima Prefecture are members.

(Kengo Hamada)

Chapter **15**

Application of Photocatalysis to Air Purification in China

Application of photocatalysis to air pollution in China

Photocatalysis is an environmentally friendly technology that originated in Japan, which has spread worldwide and formed a major industry after decades of effort since its discovery in the 1960s. Japan has led the world in commercializing and promoting the application of this technology. In the light of this trend, China has been focusing on the basic research and development of photocatalytic application products for many years. In particular, a number of studies have been carried out on the photocatalytic coating of roads to improve air pollution problems. In recent years, air pollution problems such as PM2.5 and acid rain in China have become more and more prominent, seriously threatening people's health and affecting the safety of the air environment. Nitrogen oxides, sulfur oxides (NO_X, SO_X) and VOCs emitted from vehicles and factories are the main sources of air pollution, in addition to causing urban environmental problems that threaten human health. Therefore, the removal of NO_X and SO_X is a very important issue for China.

Titanium dioxide (TiO_2) photocatalysts are very effective in removing NO_X and SO_X from the atmosphere. Under light irradiation, TiO_2 oxidizes and removes NO_X and SO_X from the air into nitrate and sulphate ions (NO_3^- and SO_4^{2-}) respectively. It is believed that this method can effectively solve air pollution problems in densely populated industrial areas and closed environments with high vehicle densities (e.g. underground car parks, motorway toll booths and service areas).

Here we present an example of how photocatalysts have been used to demonstrate their effectiveness in tackling air pollution problems in China. In 2018 and 2020, the Institute of Science and Technology of the Chinese Academy of Sciences and Rifu Jingfeng Technology (Beijing) applied photocatalytic coatings to roads in the vicinity of Bai Ma Road in the suburbs of Beijing and Xingtai City in Hebei Province (where there is a high concentration of steel manufacturing plants) respectively (see map). The air quality at the sites was then monitored in real time by the Chinese Academy of Environmental Sciences.

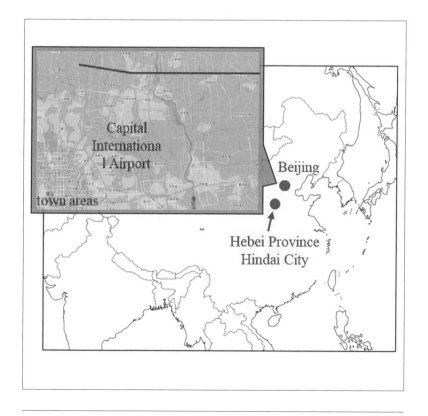

The effect of photocatalytic coating on the road near Bai Ma Road, Beijing.

The experiment was carried out twice, for four days in April 2018 and for four days in November 2018, at a road section called Bai Ma Lu in Beijing. The weather conditions at the time of sampling are shown in Table 15-1. The results of the 30-minute traffic counts on the roads are also shown in Table 15-2.

Table 15-1 Weather conditions at the time of sampling.

date(Month/day)	4/23	4/24	4/25	4/26	11/14	11/15	11/16	11/17
Temp.(℃)	23	26	28	31	12	8	8	6
weather	cloudy	cloudy	sunny	sunny	cloudy	sunny	cloudy	sunny

Table 15-2 Number of vehicles that passed through during the 30 minutes sampling period.

date(Month/day)	4/23	4/24	4/25	4/26	11/14	11/15	11/16	11/17
Large Car	115	168	175	135	133	128	155	142
Ordinary car	404	345	358	435	397	432	428	412

There were no strong winds during either of the two sampling periods; the first was during a period of relatively high temperatures and the second was during a period of relatively low temperatures. Table 15-2 also shows that traffic volumes on this road section were relatively stable.

Fig. 15-1 Sampling gas.

The air quality was monitored using a Thermo 42i-D NO_X analyzer (Thermo Fisher Scientific, detection limit 0.40 ppb) (Figure 15-1, left). The highest values of NO_2 concentration were recorded after the vehicle had passed and statistically analyzed (Figure 15-1, right).

In order to observe the trend of changes in NO_2 concentrations, two experiments were carried out. In the second monitoring, three fixed NO_2 analyzers were used to check the trend of NO_2 concentration change, while seven NO_2 sensors were used for on-line monitoring. The accuracy of the sensors ranged from 10 to 20 ppb and the sampling interval of the data was 5 seconds. Before use, the time of each sensor was adjusted to within 2 seconds of error.

Figure 15-2 shows a schematic of the test site. NO_X concentrations were measured and compared on the photocatalytically coated and uncoated road surfaces. 7 NO_2 sensor readings from C0 to C6 are shown in Figure 15-3.

Figs. 15-3 and 15-2 show that the concentration of NO_X increases along the direction of vehicle movement from C0<C1<C2. On the other hand, in the photocatalytically coated sections C3 to C6, the concentration decreases significantly and remains between 8 and 13

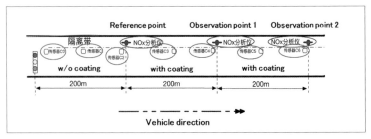

Fig. 15-2 Schematic diagram of the experimental site.
Blue circles are fixed NO$_2$ analyzers (3 locations), red circles
are NO$_2$ sensors (7 locations from C 0 to C 6).

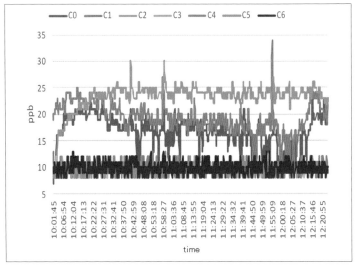

Fig. 15-3 Time variation of NO$_2$ concentration measured by 7 NO$_2$
sensors at C 0 to C 6.

ppb. The average of the measured values for each sensor is shown in Table 15-3.

The measurements from the two experiments were substituted into the above equation and the NO$_2$ decomposition rates were calculated for each and are summarized in Table 15-4.

These results show that photocatalytic coating of roads can decompose

Table 15-3 Average value of NO$_2$ concentration measured by each sensor (ppb).

w/o coating			with coating			
C0	C1	C2	C3	C4	C5	C6
16.5±3.3	19.3±2.9	23.2±2.5	9.7±0.8	9.8±1.0	9.7±0.9	9.9±1.1

$$\text{Decomposition rate}(\%) = \frac{[NO_2]_{w/o\ coating} - [NO_2]_{with\ couating}}{[NO_2]_{w/o\ coating}} \times 100$$

Table 15-4 Average decomposition rate of NO$_2$ in two experiments.

	Decomposition rate 1	Decomposition rate 2	Average
1st	14.2%	17.4%	15.8%
2nd	12.8%	14.5%	13.7%

NO$_2$ in car exhaust gases and reduce the concentration of NO$_2$ in the atmosphere.

The effect of photocatalyst coating on roads near Xingtai city, Hebei province.

Xingtai city is located in the south-central part of Hebei province and is surrounded by many power plants and steel mills. In addition to the influence of airborne pollutants from other areas, the city has been at the bottom of the air pollution ranking for many years.

In August 2020, a demonstration test was carried out by applying a photocatalytic coating to 15 roads in the vicinity of the most polluted area, the Datsukatsu spring. The area covered was approximately 25 kilometers long and 428,000 square meters. Figure 15-4 shows the changes in SO_2 and NO_2 concentrations from May before the photocatalytic coating, compared to the data from the same period in the previous year (2019).

Figure 15-4 shows that the improvement in SO_2 and NO_2 is 55% and 52.8% after the photocatalytic coating in August 2020 compared to August 2019. However, comparing the SO_2 and NO_2 concentrations in July 2020 (before the installation) with the concentrations in August 2020 (after the installation), it can be said that there is a significant reduction.

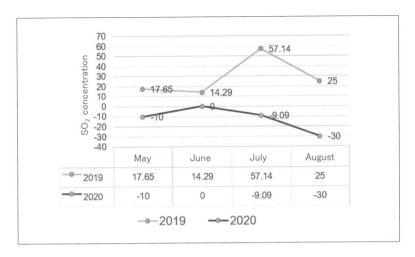

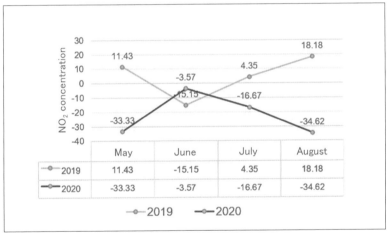

Fig. 15-4 Concentration changes of SO_2 and NO_2 before and after photocatalytic coating.

Conclusion

The photocatalytic coating on the road was found to be effective in reducing NO_2 in car exhaust and NO_2 and SO_2 around factories. Until now, the decomposition of NO_2 and SO_2 by titanium dioxide photocatalyst coating has been tested in China in small-scale or pilot test reactors. In this demonstration, the photocatalytic coating was sprayed directly onto the road to test its effectiveness. There are plans to obtain more detailed data in the future when the coating is used over a wider area.

As research progresses, a number of highly efficient, high performance photocatalytic materials are being developed, and as a result, materials with excellent performance are now available on the market. If these materials are used on the surfaces of buildings in cities, they can effectively break down pollutants in the vicinity of factories that emit car exhaust gases and NO_X. In the near future, it is expected that photocatalysts will play an active role in the field of air pollution control by controlling secondary photochemical pollution caused by NO_X.

(Jinfang Zhi)

Photocatalysis Industry Association of China

The Photocatalysis Committee of the Chinese Society of Photochemistry is an organization related to photocatalytic materials and industry in China. The main activities are as follows

(1) Jointly promote the standardization of photocatalytic materials by managing the membership system.

(2) To systematically promote the establishment of standards for the photocatalytic industry and to lay the foundation for the development of the industry.

(3) To create opportunities for exchange among industry, academia, and researchers, and to promote the exchange of industry information.

(4) To build up international exchanges and alliances, and to provide domestic and foreign enterprises with research materials on the Chinese market.

(5) To promote the promotion of photocatalytic materials by promoting the publicity of the industry and providing information.Experts and organizations from various countries in the photocatalytic materials industry are members of the association. We sincerely welcome all friends to work together and develop the industry.

Chapter **16**

Current Status of Photocatalysis in South Korea

Photocatalysis in South Korea

Research on photocatalysis development in South Korea started around 1990 with the synthesis of titanium dioxide powder and sol on a laboratory scale, environmental purification using them, and hydrogen production via water splitting, and commercialization based on this research began to occur around 2000. During this period, 23 patents related to photocatalysis development were applied for in South Korea for 10 years from 1990, but, as applied patents increased to 150 by 2005, it was after 2000 that photocatalysis attracted social attention in South Korea. Most of the photocatalytic products commercialized during this period were photocatalytic sols, which were used to coat areas where self-cleaning or oxidative decomposition of harmful organics was required (e.g., outdoor windows or indoor walls of buildings).

Then, interest in photocatalysis increased as air purifiers using photocatalytic filters and photocatalytic wallpaper for indoor use were commercialized, and more than 30 companies related to photocatalysis were established. Under these circumstances, the Photocatalysis Research Association was established in 2002, consisting of universities, research institutes and companies involved in the research and development of photocatalysis, and the Korea Photocatalysis Association was established by related companies in 2003. The Korea Photocatalysis Association aims at revitalizing the related industries and improving people's lives by improving photocatalytic technology and creating a sound market for photocatalytic products with good performance.

In 2006, a large market was formed for the coating of photocatalytic sols in the interiors of buildings such as newly built condominiums, etc., in expectation of the removal of the causative agents of "sick building" syndrome. However, after unverified photocatalytic sol products were on the market and TV reports were broadcast that questioned the effectiveness of photocatalysis because the oxidative decomposition performance of these materials could not be fully achieved in rooms with low-intensity ultraviolet light, the photocatalytic market quickly collapsed, and many companies abandoned the photocatalytic business. For almost 10 years after that, we entered a "dark period" of photocatalysis, during which not only people in companies but also researchers in research institutes could not even use the term photocatalysis. After that, the spread of Middle East Respiratory Syndrome (MERS), and the government's announcement that the air quality in the center of cities in South Korea was the worst in the world, triggered the renewed interest in photocatalytic technology in South Korea in 2015.

Table 16-1 summarizes the photocatalytic products developed in South Korea and their first commercial (The numbers in the table refer to the South Korean photocatalysis companies and their photocatalytic products listed in the Appendix). Many photocatalytic products were developed after 2015, while companies working on photocatalysis since around 2000 are few.

Table 16-1 Photocatalytic products in South Korea and first commercial sales (*companies are listed in the Appendix).

year	1995	2000	2015	2010	2015	2020
Photocatalysis product		2003Inauguration of Korea Photocatalysis Association	2015Mars outbreak	2016CEVI research team		2018GCP Project
Material		19, 18				20, 11, 1,17
Filter		19, 18		4		
Air purifier			19	4, 2	7	22, 23
sidewalk block						5, 6, 9, 10,
Cement Products						12, 13, 14, 16
Other					15	3,8,19,21

Figure 16-1 shows photos of photocatalytic materials used in the development of many photocatalytic products by South Korean companies.

Fig. 16-1 Photocatalytic material products.

(from left to right: products of companies No. 18 and No. 19 in the appendix)

The current status of photocatalytic products in South Korea will be explained by introducing the photocatalysis-related issues in two large South Korean national projects[1,2], which again brought attention to photocatalytic technology in South Korea.

Product development for antibacterial and antiviral activity of photocatalysis

In 2003, South Korea enacted a law to improve indoor air quality in public facilities such as medical institutions, schools and nursery schools, and the standards related to infection prevention were strengthened especially for medical institutions. Therefore, Photo & Environmental Technology (P&E) has developed a large air disinfection unit specially designed for hospitals, aimed at solving the problem of nosocomial infections. As shown in Fig. 16-2 (left), this equipment consists of 5 ceramic photocatalyst filter layers, 4 rows of UV lamps, a pre-filter and an ozone-removing activated carbon filter. This large air disinfection system has been installed in hospitals and public health centers all over South Korea and tested against airborne bacteria in hospitals (right side of Fig. 16-2). Before the installation of the equipment, the airborne bacteria levels in most hospitals were well above the standard values, while the airborne bacteria levels in all hospitals were lower than the standard values after the installation, confirming its excellent sterilizing effect. Furthermore, the system got a favorable reputation, because it removed not only airborne bacteria but also the characteristic odors of the hospital. We have also confirmed that the device can maintain its performance for more than two years by simply replacing the lamp after six months of use.

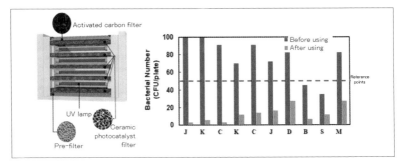

Fig. 16-2 Structure of a large photocatalytic air sterilizer and experimental results of airborne bacteria in a hospital.

(Photo: Company product in Appendix No. 19)

In the spring of 2015, 186 people in South Korea were infected with MERS, which had entered the country from the Middle East, and 38 of these infected people died. Most of the initial cases were nosocomial infections originating from a well-known South Korean hospital, which caused a huge social shock. This led to the launch of the Center for Convergent Research of Emerging Virus Infection (CEVI), a six-year, 10 billion won project led by the Korea Institute of Chemical Research (KICR) and supported by eight national institutions. The aim of the CEVI is to diagnose, prevent, cure and prevent the spread of high-risk and unspecified viruses that are likely to enter the country from abroad. The development of ultra-sensitive diagnostic techniques for viruses and preventive and therapeutic substances are being carried out by researchers from the medical field. Notably, because the results of a large air disinfection unit dedicated to hospitals were recognized, plans to use photocatalytic technology to prevent the spread of viruses were included. The aim is to use highly active photocatalysis in air purification systems for hospitals, airports and other public facilities to control the spread of viruses. Although the photocatalytic sector does not account for a large part of this project, it is significant that photocatalytic technology has been recognized as a technology to prevent the spread of viruses.

For the development of the photocatalytic air purification system, we decided to use a photocatalytic ball filter (Fig. 16-3) instead of the ceramic photocatalytic filter used in the large air disinfection system shown in Fig. 16-2. The photocatalytic ball filter is made of titanium dioxide balls mixed with titanium dioxide powder and a small amount of inorganic binder, and it was developed because it is easy to make a filter of a size that fits the irradiation range of UV-LED, which will be widely used as a light source for photocatalytic products. We produced a module using the photocatalytic ball filter and asked Kayano Sunada of KISTEC (Japan) to investigate the antiviral effect. Both influenza virus and feline calicivirus used in the test were removed in 5 minutes by more than 99% of the initial virus concentration.

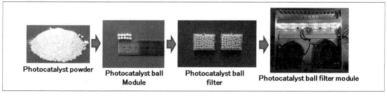

Fig. 16-3 Photocatalytic ball filter and module made with it.
(The photo shows the product of company No. 4 in the Appendix.)

Based on these results, BentechFrontier, a joint development company, has produced a household air purifier (far left in Fig. 16-4) and photocatalytic filter that can be attached to the air inlet and outlet of ordinary air conditioners or air purifiers (second from left in Fig. 16-4). The installation of photocatalytic filters in air conditioners and air purifiers in schools and other public facilities is being considered as a means of preventing COVID-19 (a novel corona virus) from being dispersed by such devices as air conditioners and is consistent with the objective of the CEVI research group to prevent the spread of the virus.

The antibacterial activity of the photocatalytic filter against *Escherichia coli*, *Pseudomonas aeruginosa* and *Staphylococcus aureus* was investigated in a large ($8 \, m^3$) experimental chamber (Fig. 16-5), which was built by the CEVI research group, and 99.9% of these bacteria were sterilized within 60 minutes.[1] We are currently applying for approval to use a virus to test the antiviral performance with this large chamber and will use the results of these experiments to develop a large photocatalytic air purification system, which is the objective of the CEVI research group.

Fig. 16-4 Air purification photocatalytic filter and photocatalytic air purifier.

(From left to right: Appendix No. 4 and No. 23 products)

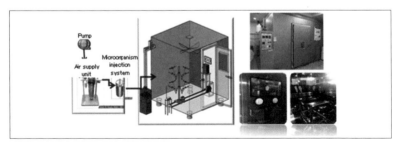

Fig. 16-5 Structure and photograph of a large experimental chamber for measuring antibacterial and antiviral performance of photocatalytic air purification system.[1]

BentechFrontier has commercialized an air purification system whose photocatalytic filter, shown in Figure 16-4, is attached to the air inlet/

outlet (center of Figure 16-4), and it is now available in South Korea. In addition, we supply the photocatalyst filter to Dewell Associates (www. del.co.jp/alkure/) in Japan, and an air purification system with this filter (second from the right in Fig. 16-4) is now available in Japan.

Chapter 16-3

Product development for air purification effects of photocatalysis

In 2018, the Green Construction by Photocatalysis Research Group (GCP) project was launched to develop construction materials using photocatalysts to improve air quality in urban areas by mainly removing nitrogen oxides among air pollutants.[2] This five-year, 16 billion won project is led by the Korea Institute of Construction Technology and involves 14 universities and research institutes in South Korea, 18 companies, as well as the Tokyo University of Science and the University of Technology, Sydney (UTS).

The following is a list of the main research issues of the GCP Project

Issue 1: Development of low-cost and high-efficiency photocatalysis production technology

Issue 2: Development of photocatalytic construction materials and application techniques for road facilities

Issue 3: Development of photocatalytic technology for housing and public facilities

Issue 4: Development of standardization techniques for photocatalytic base and construction materials

In order to use large amounts of photocatalysts in construction materials, it is necessary to secure photocatalysts with excellent performance at a low cost. Below, we introduce a new photocatalyst production technology for low-cost production of photocatalysts from sludge, jointly developed by Chonnam National University, P&E and UTS (Australia) in the first task [2,3].

Until 2013, the majority of sludge generated in the water treatment process in South Korea was disposed of by dumping into the sea. However, the dumping of sludge into the sea has been banned since 2014 due to an international treaty, and this became a major social problem, because companies unable to dispose of their sludge were forced to shut down their operations. In response, the Korean Environment Agency has been working to convert sludge into fuel and has built a sludge-fueled power plant next to a wastewater treatment plant that discharges large quantities of sludge. However, the use of chemical sludge as fuel causes a number of problems when the sludge contains high levels of metal coagulant. Therefore, we tried to solve the problem of chemical sludge treatment by producing a photocatalyst from sludge, as shown in Fig. 16-6. We developed a method to use titanium salt [$TiCl_4$, $Ti(SO_4)_2$] instead of the coagulant (e.g., Al salt and Fe salt) used in the existing coagulation process of wastewater treatment. When titanium salt is used as a flocculant, the chemical sludge contains titanium components, and only titanium dioxide remains when the organic matter is burned. Laboratory and pilot tests of this method on a

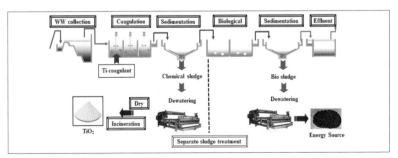

Fig. 16-6 Wastewater treatment using titanium salt coagulant and production of titanium dioxide from sludge[2,3].

range of wastewater and sewage applications, as well as field trials, have shown, in collaboration with researchers in Australia and China, that this technology is highly effective in the treatment of wastewater containing dyes.

In the production of titanium dioxide from chemical sludge, at calcination temperatures of 600-650° C, titanium dioxide with anatase phase is formed, while this phase converts to rutile at calcination temperatures above 800° C. The size of the titanium dioxide particles obtained by calcining the chemical sludge at 600° C was about 20 nm, estimated from TEM photographs, and the BET surface area measured from nitrogen adsorption experiments was slightly larger than that of the well-known P-25 TiO_2 powder. The photocatalytic activity was investigated from the acetaldehyde decomposition and was found to be comparable to that of P-25. Field experiments using this technology at a wastewater treatment plant in a dyeing industrial complex in South Korea (20,000 tonnes per day wastewater treatment scale) confirmed that the production of about 10,000 tonnes of photocatalytic titanium dioxide per year is possible. If this technology can be established, it is expected that photocatalysts will be widely used in areas where their use has been difficult due to their high price. This technology won the Global Grand Honor Award in the Applied Research category of the International Water Association's 2012 IWA Global Project Innovation Awards. We have published more than 70 papers and registered patents in South Korea, USA and China.

Fig. 16-7 shows the photocatalytic cement products as construction materials produced by the companies in the second project of GCP.

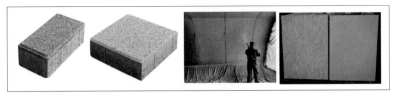

Fig. 16-7 Photocatalytic cement products.

(from left to right: products of Appendix No. 10, No. 12, and No. 13)

When we developed this technology, we expected to obtain visible light-responsive titanium dioxide doped with carbon or nitrogen by calcination of the sludge, because the sludge contains a large amount of organic matter and nitrogen. However, the titanium dioxide obtained did not show such high visible light response.

Therefore, we tried to develop a visible light responsive photocatalyst by calcinating melamine, which contains a high nitrogen content, together with titanium dioxide (NP-400) produced from sludge.[4] The calcination of NP-400 and melamine was conducted in an atmospheric furnace at a rate of 10° C/min up to 550° C for 3 hours. This resulted in the formation of a graphitic carbon nitride (g-CN)/TiO$_2$ composite (TC), and, by changing the amount of melamine, TC-X with different ratios of g-CN and TiO$_2$ were produced. The composite of g-CN and TiO$_2$ was observed from the TEM photograph of the TC-4 sample in Fig. 16-8 (left). Fig. 16-8 (middle) shows the UV-visible absorption spectrum of NP-400, g-CN and TC-X. NP-400 cannot absorb visible light longer than 400 nm, while some TC-X samples can absorb up to 600 nm, depending on the amount of melamine. The results of NO removal experiments with NP-400, g-CN and TC-X under visible light conditions (Fig. 16-8 (right)) show lower photocatalytic performance in comparison to those of NP-400 and g-CN, while the photocatalytic performance of the TC-3 and TC-4 samples was high. These results confirm that visible light-responsive photocatalysts can be produced by sintering titanium dioxide produced from sludge with an appropriate amount of melamine. In the future, we plan to further develop these synthesis methods and investigate the use of waste materials such as melamine resin.

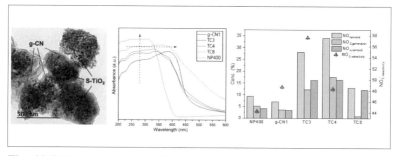

Fig. 16-8 Physical properties and photocatalytic activity of NP-400, g-CN and TC-X under visible light irradiation.[4]

One of the photocatalytic products targeted for development in this project is a photocatalytic paint. The removal of NO_X is important in improving air quality in the city center, and we believe that the application of photocatalytic paint to the exterior walls of buildings will be an effective way in removing NO_X. Because the main application of photocatalytic paints currently on the market is to prevent staining of buildings due to the self-cleaning function of the photocatalysis, their performance in NO_X removal is not as high as expected. This implies that it is difficult to produce a photocatalysis paint with both high photocatalytic performance and durability, which is the original purpose of the paint.

In collaboration with Inha University, P&E and SH Seoul Housing City Corporation, which manages many high-rise condominiums in Seoul, we have developed a photocatalytic paint with high NO_X removal performance while maintaining durability for use on building exteriors.[5] The photocatalyst (INCN) mixed into the photocatalytic paint was prepared by chemical treatment of the base photocatalyst (INHS) endowing it with both hydrophilic and hydrophobic properties, as shown in Fig. 16-9 (left). In the aqueous paint containing chemically treated INHS, we expect that a large amount of INHS is located on the surface of the paint, as shown in Fig. 16-9 (right).

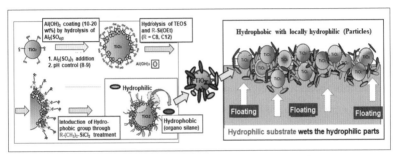

Fig. 16-9 Schematic diagram of INCN photocatalyst production method and water-based paint mixed with INCN photocatalyst.

The NO$_X$ removal performance of photocatalyst paint prepared by mixing aqueous paints with 10 and 20 wt% of untreated or surface-treated INCN photocatalyst was investigated. The NO removal performance of the aqueous paints with INHS was low, and even though the amount of INHS increased from 10 wt% to 20 wt%, the NO removal performance remained almost the same, i.e., 1.3% and 1.4%, respectively. We speculate that this is because the aqueous paint largely covers the INHS photocatalyst. On the other hand, the NO removal performance of the aqueous paint with 10 wt% of chemically treated photocatalyst (INCN) was 7.3%, and became significantly higher (18.2%) when the INCN content was increased to 20 wt%. This work was supported by the Photocatalysis International Research Center,

Research Institute for Science and Technology, Tokyo University of Science, and we thank Akira Fujishima, Director of the Center, and Norihiro Suzuki, Junior Associate Professor, for the opportunity to collaborate.

The durability of photocatalytic paints containing 20 wt% each of INHS and INCN was investigated by two weeks of UV irradiation at 350 W/m^2, which is more severe than the experimental conditions in a general durability test. In the case of INHS-20, a large amount of powder was observed on the surface after 2 weeks of UV exposure, whereas, in the case of INCN-20, there was almost no change in the surface after 2 weeks. This indicates that aqueous paint with chemically treated photocatalyst (INCN) photocatalyst has no durability problems. Therefore, SH Seoul Housing and Urban Development Corporation reported that the photocatalytic paint with 20 wt% of INCN photocatalyst meets the product standard for photocatalytic performance and durability set by the Corporation.

Summary

As an overview of the current status of photocatalysis in South Korea, we introduced the topics related to photocatalysis in South Korean national project and explained that product development is being promoted with focus on their antibacterial and antiviral functions for indoor use, and air purification functions such as NO_X removal for outdoor use. In particular, the fact that photocatalytic products, which were approved as a MERS countermeasure in 2015 and started to be used to prevent the spread of viruses, were actually used in public health centers in South Korea during the current outbreak of the novel coronavirus is a great achievement for the developers. We believe that the antiviral functions of photocatalysis will be a powerful tool to protect human life and health from the continuing threat of viruses in the coming years.

We believe that it is now necessary to develop a new structure of air purifier, not only to install photocatalytic filters in air purifiers. A humidifier with a photocatalytic filter for water purification would be a good solution. Fig. 16-10 shows a photocatalytic filter for water purification and a water purifier using this filter. It is also necessary to develop indoor interior products that can maximize the antiviral function of photocatalysis even under indoor light conditions.

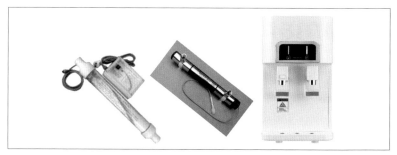

Fig. 16-10 Photocatalytic filter and water purifier products for water treatment.

(From left to right: Appendix No. 19 and No. 15 products).

Recently, even though public interest in photocatalysis has become very high in South Korea, there is no fair organization to evaluate and certify the standards and performance of commercial photocatalytic materials or products. To meet the expectations of society, we would like the Korea Photocatalysis Association, established in 2003, to play this role; however, since the Association is an organization of related companies, its role seems to be limited. In order to fulfill the original purpose of the Association, which is to create a sound market for reliable photocatalytic products and to contribute to the revitalization of the photocatalytic industry and improvement of people's lives, the cooperation of not only the Association and related companies, but also the government, which supervises the Association, is necessary.

References

1. Project Plan and Results Report of the Center for Convergent Research of Emerging Virus Infection (CEVI)

2. Green Construction by Photocatalysis Research Group (GCP), Ministry of Land, Infrastructure, Transport and Tourism, South Korea.

3. Shon, H.K.; Vigneswaran, S.; Kim, In S.; Cho, J.; Kim, G.-J.; Kim, J.B.; Kim, J.-H., Preparation of titanium dioxide (TiO_2) from sludge produced by titanium tetrachloride $(TiCl_4)$ flocculation of wastewater, Environmental Science & Technology 2007, 41(4), 1372-1377.

4.

4. Hossain, S.M.; Park, H.; Kang, H.-J.; Mun, J.S.; Tijing, L.; Rhee, I.; Kim, J.-H.; Jun, Y.-S.; Shon, H.K., Facile synthesis and characterization of anatase TiO_2/g-C_3N_4 composites for enhanced photoactivity under UV-visible spectrum, Chemosphere 2020, 262, 128004. 5.

5. Kim, J.-H.; Hossain, S.M.; Kang, H.-J.; Park, H.; Tijing, L.; Park, G.W.; Suzuki, N.; Fujishima, A.; Jun, Y.-S.; Shon, H.K.; Kim, G.-J. Hydrophilic/ Hydrophobic Silane Grafting on TiO_2 Nanoparticles: Photocatalytic Paint for Atmospheric Cleaning. Catalysts 2021, 11, 193.

Appendix

List of Photocatalyst Companies and Photocatalyst Products in South Korea

No	Company name	Photocatalysis products	HP
1	Airmarah Co Ltd	Photocatalytic sol	www.airmarah.com
2	APCTEC Co Ltd	Photocatalytic air purifier	www.apctec.co.kr
3	APEC Co Ltd	Photocatalytic Paint	www.a-pec.co.kr
4	BentechFrontier Co Ltd	Photocatalytic Filter Air Purifier	www.btfgreen.com
5	decopave Co Ltd	Photocatalysis Walkway Block	www.decopave.co.kr
6	DESIGN BLOCK Co Ltd	Photocatalysis Walkway Block	www.dbw.kr
7	DAEHEUNG MITAL Co Ltd	Photocatalysis Air Purifier	www.dhmt.co.kr
8	Dongnam Co Ltd	Photocatalytic Sol	www.dongnamad.co.kr
9	EDC LIFE Co Ltd	Photocatalytic Cement	www.edclife.co.kr
10	Get-pc Co Ltd	Photocatalysis Walkway Block Photocatalytic Pannel	www.greencon.org
11	J-chem Co Ltd	Photocatalysis Walkway Block	www.j-chem.net
12	JH Co Ltd	Photocatalytic Cement	www.jh-corp.co.kr
13	JH Energy Co Ltd	Photocatalytic Pannel	www.jh-corp.co.kr
14	JW Co Ltd	Photocatalysis Walkway Block	www.jeongwoo.co.kr
15	HANVIT KOREA Co Ltd	Photocatalytic Water Purifier	
16	MBLOCK Co Ltd	Photocatalysis Walkway Block	www.m-block.kr
17	Nano M Co Ltd	Photocatalytic sol	www.nano-m.com
18	NANOPAC Co Ltd	Photocatalytic materials Photocatalytic filter	www.nano-pac.com
19	P&E Co Ltd	Photocatalytic materials Photocatalytic filter	www.pnekr.com
20	P.T KOREA Co Ltd	Photocatalytic road coating	www.paveteckorea.com
21	scienceceramickorea Co Ltd	Photocatalytic Ceramic Products	www.scienceceramic.com
22	Sewon Co Ltd	Photocatalytic Air Purification System	www.sewoncentury.co.kr
23	YUNSUNG Co Ltd	Photocatalytic Air Purification System	www.yunsungcompany.com

(Jong-ho Kim)

Chapter **17**

Status of Photocatalysis in Europe

Global situation about commercialization and standardization of photocatalytic technologies in Europe

Industrial situation for photocatalysis in Europe ?

Global situation about commercialization and standardization of photocatalytic technologies in Europe

The present author became involved with the application field in Europe from the early beginning in initiating the manufacture of photocatalytic ceramic tiles Hydrotect (TOTO–RAKO–DSCB project) in 1997 and promoting novel commercial applications (including the protection of world culture heritage). This contribution is based on existing facts as well as on discussion with top European experts in the commercial field but shall be still considered partly as a personal view of its author. It is very difficult to collect all reliable data about all of the present commercial activities in Europe. Please accept our apologies that not all of the important players active commercially in Europe are mentioned. On the other hand, trends of photocatalytic business in the future are in agreement with EU policy.

Introduction

Photocatalysis fundamental research in Europe has a long, almost 100-year tradition. In contrast, the commercial application field is much younger and perhaps was accelerated by the discovery of the hydrophilicity of photoactive material surfaces much later. Italcementi, a global cement manufacturer, can be noted as introducing photocatalytic cement with self-cleaning and antipollution functions in Europe. The Italian Pavilion at the last World Expo, hosted by Milan in 2015, was constructed from very unique photocatalytic self-cleaning concrete parts

Fig. 17-1 The Italian pavilion at the World Expo in Milan in 2015, on which a photocatalyst was used.

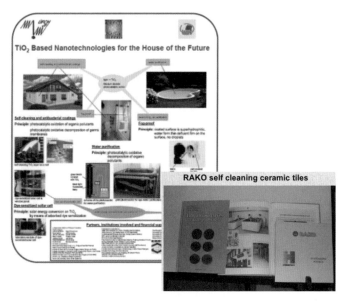

Fig. 17-2

(Top) A Czech company advertised the great potential of photocatalysis extensively (2005).
(Bottom) RAKO-Hydrotect tiles are introduced in the Czech version of "Fundamentals and Applications of Titanium Dioxide Photocatalysis."

made by that company. Czech ceramic tile manufacturer RAKO Rakovník (daughter company of German DSCB Co. at that time), in cooperation with TOTO in Japan may be also mentioned as one of companies pioneering commercial applications in Europe. Photocatalytic, antibacterial and self-cleaning Hydrotect tiles appeared on the market first in 1999. Generally, it can be concluded that the progress of successful commercial applications in Europe has been tied to the development of that field in Japan.

Titanium dioxide as fundamental photocatalytic material

Almost all photocatalyst material presently used for commercial activity in Europe is based on TiO_2. There are several major TiO_2 manufacturers located all over Europe, and most of them have photocatalytic TiO_2 forms in their production portfolio. Manufacturers of TiO_2 located in Europe include the following:

Cinkara – Slovenija

Cristal (Tronox) – France, UK

Huntsman –Finland, France, Germany, Italy, Spain, UK

Kronos –Belgium, Germany, Norway

Precheza –Czech Republic

TRONOXLLC - Holland

Police –Poland

Unfortunately, after a long classification process, the European Union (EU) decided to classify titanium oxide as a suspected carcinogen (category 2) by inhalation in certain powder forms (applied to the whole group of products with TiO_2) on 18 February 2020. This will apply on 1 October 2021 (after a 18-month transition period). For photocatalytic applications, this classification could lead to a serious acceptance problem, which can slow down this still struggling technology in Europe.

Photocatalytic product standardization and certification

Similar to the Japanese experience, the first commercial applications in Europe were leading to the conclusion that successful commercialization of photocatalytic products is strongly dependent on the development of testing methods accepted at the ISO or CEN level to guarantee the advertised function. Establishing CEN standardization system was one of the main objectives of the European COST 540 (Cooperation in Science and Technology) project "Photocatalytic technologies and novel nanosurfaces, materials, critical issues." COST 540 brought together experts from 20 Europeans countries with Japanese colleagues from PIAJ invited as guests to join collaborative work. Focusing on the development of photocatalytic (nano) materials for practical applications, together with activities leading to the establishment of the CEN Technical Committee for Photocatalysis may be mentioned as the most important goals of the COST project.

The first CEN/TC 386 meeting was organized in France (Paris) in November 2008 with the AFNOR (French standard association) committee secretariat. The meeting in 2020 and 2021 were organized, because of the COVID-19 pandemic situation, via videoconference (Zoom) only, as the 13th and 14th Plenary Meetings. In comparison with ISO, CEN/TC for photocatalysis is independent. Photocatalysis in

ISO is organized within ISO/TC 206 for fine ceramics as WG 9. This fact creates a slight legislative problem in ISO-CEN relations but is not influential for the real collaborative work. e.g., CEN/TC 386 decided to review ISO/CD standard 24448 –LED light source for testing semiconducting material used under indoor lighting environment during the 13th Plenary Meeting because of the active participation of a CEN/TC expert, helping significantly to establish the ISO standard.

Standard's development process
Detailed Timeframe

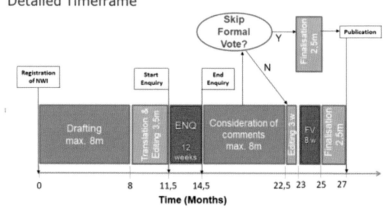

Fig. 17-3 Time schedule of the standardization process.

The philosophies of the CEN and ISO approaches to standardization in photocatalysis seem to be similar. Within CEN/TC 386, there were originally 8 working groups (WGs) established:

WG1 "Terminology" (Convenorship Dr. Claudio Minero / Italy)

WG2 "Air-purification" (Convenorship Dr. Chantal Guillard / France)

WG3 "Water-purification" (Convenorship Dr. Anastasia Hiskis / Greece)

WG4 "Self-cleaning applications" (Convenorship Dr. Claudio Minero / Italy)

WG5 ("Medical applications") – WG is closed

WG6 "Light sources" (Convenorship Dr. Zissis / Greece)

WG7 "New technologies and other important issues" (Convenorship Dr. František Peterka / Czech Republic)

WG8 "Microbiology" –

In reality, the WG 5 group was never active and soon closed. WG 8 is very important today, because of COVID-19, but has problems due to a lack of experts and is rather dormant. WG 3 activities are limited because of only limited recent commercial applications. The EN 17210 standard "Performance of photocatalytic materials by measurement of phenol degradation" is already in existence. Further development of new standards within WG 5 is slowing down. The CEN approach to self-cleaning is based on solid-to-solid contact, which is different from the ISO concept. The EN 16845-1:2017 standard "Photocatalysis – Anti-soiling chemical activity using adsorbed organics under solid/solid conditions - Part 1: Dyes" demonstrates this.

CEN standards are already in existence for the most important applications of NO_x removal and self-cleaning. Several proposals having the status NWI (New Working Items), and several standard items received the status TR (Technical report). The number of already approved CEN standards is still slightly behind that of approved ISO standards because of a later start. Surprisingly, the fact that there have been more experts from different European countries in CEN/TC 386 WGs in comparison with those under ISO/TC WG 9 is also slowing down the evaluation process. The large number of different opinions on proposed standards often result in endless, strictly opinionated discussions. The already approved ISO standard "Photocatalytic activity indicator ink" originally proposed as a CEN standard can serve as an example. After the concept of the ink standard being rejected by several European experts, the proposal was withdrawn from CEN/TC.

In additional to the ISO standardization system, CEN/TC 386 approved a standard regarding the testing of photocatalytic air cleaners. EN 16846-2 "Measurement of efficiency of photocatalytic air cleaners used for the elimination of VOC and odors in indoor air – Part 2: Tests in large chamber" proved to be very effective in distinguishing properly designed photocatalytic devices from fakes. The most important part of the test is the measurement of CO_2 released from the decomposed mixture of organic pollutants; the correct theoretical amount of CO_2 released can easily be calculated. The absence of CO_2 release means that an air purifier is using other technology, e.g., absorption, while a high CO_2 release means that there is an error in the device construction.

WG 7 of CEN TC386 is still working on the concept for "Photocatalytic material ageing and durability of photocatalytic materials." Different opinions of experts from ISO and CEN shall be harmonized to become fully standard. Work already completed on the issue is published as Technical Report (TR) at the moment.

Another important CEN standard has been proposed by Spain for in situ measurement of the photocatalytic activity of construction materials. The new standard proposal was one of the important results of the LIFE –PHOTOSCALING project.

CEN/TC 386 has several liaisons with CEN/TC 352, "Nanotechnologies" and ISO/TC 206/WG 9 "Fine ceramics - Photocatalysis" being the most important. The number of European laboratories capable of providing all existing CEN standard tests is unfortunately limited, mainly for economic reasons. There are limited inquiries for testing, because companies active in Europe are not forced to test their products by standard methods.

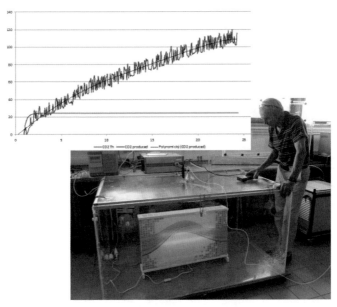

Fig. 17-4 EN 16846-2 test (at right is the author *of this chapter*).

In any case, the public trust in the performance of photocatalytic material technology is the key issue to succeed, e.g., poor performance of photocatalytic surfaces for NO_X removal is often caused by poor materials. Japan and PIAJ realized that product certification must follow standardization. To distinguish between good and poor performance for advertised functions, PIAJ introduced a labelling system based on passing the existing ISO tests with good results. CAAP, in cooperation with PIAJ, next introduced the CAAP Certification system, which is offered to companies in the Czech Republic, as well as to other companies in Europe. The system is valid generally. In contrast to the PIAJ label, for obtaining a CAAP Certificate, membership in the Photocatalytic Association is not the rule. For photocatalytic materials applied to NO_X removal, FAP also introduced a Voluntary Certificate based on passing the ISO standard with very good results. Recognition and acceptance of this kind of Certificate in the commercial field is still under broad discussion.

Successful photocatalytic applications in Europe and organizations promoting commercial application fields

The first European companies introduced photocatalytic products on the market round 2000. Commercial business in Europe was far behind the "light cleaning revolution" in Japan at that time. In order to speed the commercial activities in Europe, the European–Japanese Initiative for Photocatalytic Application (EJIPAC) was established). After the EJIPAC stopped its activities in 2008, some European countries started to follow Japan's example. The Photocatalytic Industry Association of Japan (PIAJ) was established in 2006, and France initiated the establishment of a photocatalytic federation on the European level. The European Photocatalytic Federation (EPF) was born in 2009. Independent Czech (CAAP), German (FAP – organized under the German Association of Producers of Pigments and Fillers) and Spain (Iberian) federations followed later. Czech, French, German, Iberia committees existed originally within the EPF structure, with other members from Belgium, Italy, the United Kingdom being most active. EPF brought together more than hundred European members, including major players in the photocatalytic field and large international companies (e.g., Saint Gobain), several scientific organizations and individual experts at their peak in 2013. After the EPF decision to cease its activities in 2017, European national federations and their cooperation with PIAJ become very important to further promote photocatalytic business in Europe. For the commercial application field in Europe, after the EPF termination, meetings of experts within CEN/TC 386 become very helpful to keep cooperation alive.

FAP, with 18 members, known international companies (BASF, Heidelberg Cement) included, and CAAP, with 12 members (with number 1 Czech paint manufacturer BAL), extended their contacts after 2017, especially because of the problem of TiO_2 toxicity and the important task of defending the photocatalytic business in Europe.

Similar to Japan and worldwide, it is possible to divide the commercial applications in Europe into six groups:

1. Environmental control in urban zones (roads, buildings, roofs, etc.)
2. Self-cleaning surfaces, including self-cleaning with strong anti-microbiological effect
3. Interior air treatment (air cleaning devices)
4. Water purification
5. Energy conversion
6. New trends, novel applications

Actual business activities are covered by the first three groups only. According to a BCC study, Europe was covering 25% of photocatalytic commercial applications (2015). Environmental control activities as the most important are described below in more detail.

Environmental control in urban zones--air cleaning

Air cleaning applications have become perhaps the most important target in the photocatalytic applications buildup because of the EU strict regulations on emission control. To keep pollutants (specially NO_X) below limits, reliable and effective technologies or measures have been sought after. For those, large European companies, including TiO_2

photocatalyst, cement, roofing, façade material, flooring manufacturers were focused on that goal, trying to develop photocatalytic products accordingly. For example, Italcementi devoted much effort to succeed with the "photo-road" idea. EU and national governmental organizations financially supported at least five joint European pilot projects to demonstrate the efficiency of photocatalysis for NO_X reduction under real conditions. The National Life PHOTOSCALING project, supported by the Madrid city government, which was completed in 2019, may be the most recent one. The author of this contribution was deeply involved in this. His patented photoactive nanocomposite material was tested during one of the projects for pollution control in Holland. He was also one of evaluators for the PHOTOSCALING project.

The results of that investigation may be summarized as follows:

The photocatalytic performance for NO_X reduction was demonstrated for the good and excellent photocatalytic materials. During testing, almost all existing photocatalytic materials available on the European market were tested with poor or good performance Thus, photocatalytic material certification is very important to identify good materials for a required task. The NO_X reduction can be very high at short distances from the treated surfaces such as roads or paving stones, roofing, etc. (30-60 %) and high in narrow "canyon" streets (15-30 %). Realistic NO_X reduction for the whole city can be estimated only. It is strongly dependent on the realistic number of surfaces that can be treated with photocatalysts, as well as weather conditions. Both are playing most important roles.

Photocatalytic products as roofing materials (ICOPAL, Erlus), ceramic tiles (Rako, TOTO Europe), felts, painting and coating systems for façades, plasters (BAL, Sto), concrete elements (Italcementi) are the most common. Glass has also been an important target (Saint Gobain, Pilkington).

To defend the photocatalytic material application for air pollution control and acceptance by EU governmental organizations, it is necessary to prove that there is:

— No significant side reaction forming toxic NO_2

— No significant organic binder degradation leading to formaldehyde

Environmental control indoors is another story, because modified photocatalysts for indoor condition performance are rather scarce, and to apply artificial UV-A sources in interiors is often not practical. However, there is a market in Europe, which is often misused by the marketing skill of imported bogus products and technologies, killing trust in photocatalysis.

Air cleaning devices have proved to be more effective for interior air cleaning, and there are several European manufactures of air cleaners at present. During the COVID-19 pandemic time, it is clear that there is huge market for good air cleaning systems of this type. The photocatalytic and other AOT (Advanced Oxidation Technology) types are becoming popular.

Reliable TiO_2 photocatalytic material manufacturers with the good marketing strategy can still be successful with this very important application, e.g., large international companies such as TRONOX, new owner of Crystal TiO_2 manufacturer, same as PHOTOCAT, or Heidelberg Cement, which purchased Italcementi.

Self-cleaning, with easy evidence of the effectiveness of this application, has become reliable, but the market for esthetic function only is limited. However, several European companies are still selling their products as self-cleaning with anti-pollution function as a bonus. e.g., Czech BAL Co., German Sto Co., and several companies in Spain).

Self-cleaning with antimicrobiological effects (bacteria, viruses, algae, fungi, moss) and applications focused on surface temperature control related to the

Albedo effect are becoming more important. With environmental changes, new technologies, and the pandemic situation, photocatalysis in Europe is going through trials of possible commercial applications in this field. Perhaps Czech companies such as BAL–Teluria with Nanotec System are pioneering this application field. Behind this is the fact that, in the Czech Republic, there are thousands living in houses that are additionally thermally insulated, which are later attacked by algae and fungi. Removal by photocatalytic action is more ecologically friendly than the short-term function of chemical substances.

Important dates of commercial applications in Europe (personal view of the author of this contribution)

1995-2000 Start of the one of first photocatalytic commercial applications in Europe. For example, RAKO Rakovník (member of DSCB group) started to manufacture photocatalytic antibacterial tiles Hydrotect in 1999 under TOTO license.

2000-2005 More European companies discovered the potential of photocatalytic applications and introduced several photocatalytic products as paints, concrete, glass, air purifiers on the market, based on in-house technologies or those under Japanese licenses (Sto, BAL, DSCB, Italcementi and many others).

2001 First standards on the national level in order to properly evaluate photocatalytic function were introduced as UNI (Italy) or DIN (Germany). Later, after ISO/TC 206 was established, ISO standards were utilized. First approved ISO standard evaluating function for NO_X removal in 2007 is still generally very popular.

2005-2009 European COST 540 project focused on new materials development, application and standardization (15 seminars with Japanese participation, supporting reliable applications and standardization, were organized in various European countries).

2004-2008 EJIPAC (European-Japanese Initiative for Photocatalytic

Fig. 17-5 Exterior walls of houses and historical buildings with algae and fungal growth.

Application Commercialization) was organizing conferences aiming to extend cooperation on a commercial basis between Europe and Japan.

2007 The RILEM conference was organized in Florence, Italy by Italcementi with the aim to promote photocatalytic applications in the construction industry.

2008 CEN/TC 386 for Photocatalysis was established to allow more European national representatives to create standards for commercial business in Europe

2009-2017 EPF (European Photocatalytic Federation) was born by extension by French initiative. EPF attracted more than 100 members during the photocatalytic application boom. EPF issued the "White Book of Photocatalysis" in 2015, among others After several important members left, EPF decided in 2017 to terminate its activities. In the meantime, national federations were created, e.g., CAAP (Czech Federation for Applied Photocatalysis) in 2013.

2014 An international meeting focused on Standardization and Certification helping Commercial Application was organized in Praha. Representatives of existing Photocatalytic Associations from Europe and Japan were presenting achievements and expectations, as well as problems in the commercial application field.

2014 CAAP–PIAJ signed a cooperation agreement helping to introduce Photocatalytic Product Certification in Czech Republic, based on the PIAJ labelling system, having general validity in other EU countries.

2017 CAAP started to issue Certificates proving good and excellent function for existing photocatalytic applications.

2017 A FAP–CAAP meeting focused on strengthening cooperation in Europe on photocatalysis after EPF termination. Rejecting the TiO_2 toxicity classification in the EU was an important task of the meeting.

2017 European experts contributed to the Japanese publication "Photocatalytic World," showing the potential of commercial applications in Europe, among others.

2020 Photocatalysis is still playing an important role as a prospective technology in the EU. Several European companies are commercially successful, especially in the field of environmental catalysis, self-cleaning surfaces and air cleaning purification devices for interiors. Novel applications focused on the prevention of microbiological pollution, surface overheating proved to have commercial potential and good prospects.

Future and trends of commercial applications in Europe

Companies in Europe having commercial business in the photocatalytic field defend the important position of photocatalysis in environmental and self-cleaning applications, in spite of the fact that the toxicity classification of TiO_2 in early 2020 is already resulting in problems for several of them. In particular, large international companies located in Europe need to explain well their application use of TiO_2. EU strategy and priorities are also changed and European legislation is focused more on renewable energies and decarbonized chemistry in the first place. Decreasing NO_X pollutant levels is still considered important, but reducing traffic with conventional engine-vehicles as a major source of NO_X is seen as a solution for the long-term horizon. However, recently, photocatalysis remains as an efficient active method to control emission/pollution in cities and suburban areas. In any case, only novel challenging applications by taking advantage of "light cleaning" by the sun's energy is a guaranteed photocatalytic application field. These may be connected with novel methods for the protection of culture heritage or preventing surface overheating. This direction, by use of a nanocomposite system with photoactive function, is followed by Czech companies (BAL, PUR, Nanotec System) for example. Photocatalytic methods will also provide options for the de-centralized production of synthetic fuels and other valuable chemicals using carbon dioxide in the future.

The killer phrase 'what happens in the night or in the shade?' is hard to avoid but cannot stop us.

(František Peterka, Ph.D., Chairman of all European COST project

Photocatalytic technologies)

Chapter 17-2

Industrial situation for photocatalysis in Europe

The commercial market for photocatalytic products in Europe is increasing. The introduction of photocatalytic products in Europe twenty years ago opened a market for smart materials. However, after some years of stagnation, the overall revenue on sold and installed photocatalytic products has increased during the last five years.

In Europe, we face a change in the concept of using photocatalysis. It all began with self-cleaning surfaces and photocatalysis to improve and maintain buildings' aesthetics to keep them cleaner over time. Then it changed towards NO_x-removing properties and to prove the effect with standards and real-life testing. The recent research & development trend is towards anti-organic growth on surfaces such as anti-algae and anti-viral effects. We foresee that, for photocatalysis in the near, we will see companies combine the properties of photocatalysis into a sustainable concept with a social impact, including environmental, climate, social, and economic effects.

The main category of products sold today with photocatalytic effect are products for the building industry, e.g., pavement, facades, roads, and roofing. The largest volume of photocatalytic products sold is in concrete building materials – pavement and facades – and paint producers. The leading players within the photocatalytic industry in Europe are within these product segments.

The main driver for selling and promoting photocatalytic products is the air purification properties to reduce NO_x pollution in the urban environment. Many products are launched with NO_x-reducing properties, and millions of installed square meters purify the air we breathe every day.

Photocatalysis in Europe

The photocatalytic technology came to Europe from Japan. The first products on the market were in the range of self-cleaning surfaces produced by major concrete and cement producers and window glass manufacturers like Pilkington and Saint-Gobain, where the glass surface was rendered self-cleaning during the manufacturing process. From 1990 to 2000, the main focus was on self-cleaning properties to reduce maintenance cost and improve surfaces' aesthetic look by keeping the surface cleaner over time.

After 2000, the main driver for photocatalysis in Europe was demonstrating the air-purifying effect of different photocatalytic products. Since many products were introduced in a short period with various product claims, the different standards for testing photocatalytic products were introduced. Several working groups were established to improve the way of testing photocatalytic products by certified laboratories. The primary purpose was to set a common photocatalytic standard to characterize different products and compare the NO_x-reducing properties.

The most widely recognized test procedure applied to determine the activity of NO_x-removing building materials using photocatalysis is the ISO 22197-1 standard test. This standard utilizes the flow-through method, in which 1 ppm NO gas passes over the sample while it is illuminated by a UV lamp with an intensity of 10 W/m2 in the wavelength range 300-400 nm. The NO_x concentration is measured at the inlet and outlet. To allow comparability of different studies, tests should be conducted using the same procedure, preferably a standard test such as the ISO 22197-1 method. Other test methods include the UNI and CEN methods, among others.

The photocatalytic associations in Europe were the main drivers for developing and introducing the standard test methods.

Real-life testing of photocatalytic products for NO$_X$ reduction

In the years from 2005 to 2010, the first real-life tests were performed in Europe. From 2005 and forward, several different EU-funded projects were initiated to prove the effectiveness of photocatalytic technology in real life. The main challenge was to create a protocol on how to test and validate the real-life effect of NO$_X$-removing photocatalytic surfaces. Monitoring NO$_X$ levels in urban space is challenging, and the aim to also detect and validate changes due to photocatalytic effect made it even more difficult.

A recent review article (published in the Journal of Photocatalysis 2021)[1] examines the real-life tests performed from 2005-2020. The authors propose a general scheme to evaluate the quality of performed real-life tests, but the evaluation criteria can also be used to design new real-life tests. Pedersen et al. focused on the use of TiO$_2$ as a photocatalyst in NO$_X$-removing construction materials, which in a recent report by Environmental Industries Commission (EIC) was highlighted as one of the cheapest options for NO$_X$ removal.[2]

In the review, the aim was to objectively evaluate the quality of the existing field studies from 2005 to 2020. The review concludes that TiO$_2$-based photocatalysis can remove hazardous air pollutants, e.g., NO$_X$. Incorporating this technology into surfaces in urban areas, such as pavement, asphalt, tunnel walls, etc., is a promising tool to improve air quality at a low cost in areas where pollution levels and population densities are high. Quantifying the

Study[a]	Lab	Area	Distance	Ref.	Blank	Duration	Frequency	Durability	Suppl.
Antwerp	★	★	-	★	-	★	-	★	★
Guerville	-	-	★	★	-	★	★	-	★
London	-	-	★	-	★	★	★	-	★
Rome	★	★	★	-	★	-	★	-	★
Hengelo	★	★	★	★	★	★	-	★	★
G.I.T	★	-	-	-	-	-	-	-	-
Malmø	-	★	★	★	★	★	★	-	★
Manila	-	★	-	-	★	★	-	-	★
Brussels	★	★	-	★	★	★	★	★	★
Louisiana	★	-	-	-	★	-	★	★	★
Wijnegem	★	-	-	-	-	★	-	★	-
Gasværksvej	★	★	★	★	★	★	★	-	★
CPH airport	★	★	★	★	★	★	★	-	-
The Hague	★	★	-	-	-	-	★	-	★
Fælledvej	-	★	★	-	★	★	★	-	★
Holbæk mv	-	-	-	★	-	★	★	-	★
Valencia	-	★	★	-	-	-	★	-	-
Toronto	-	-	-	-	-	★	★	-	-
Roskilde	★	★	-	-	-	★	-	★	-
Putten	-	-	-	-	-	-	-	-	★
Tsitsihar	★	★	★	★	-	★	★	-	★
Madrid	★	★	★	★	★	★	★	★	★

[a] References for the studies are given in the main text

Fig. 17-6 Overview of 22 real-life tests and 9 evaluation criteria.[1]

actual reductions in real life is challenging due to the many parameters affecting the test results, such as weather conditions and traffic levels, and real-life studies with conflicting results are presented in the literature. In the review, the authors attempted to evaluate 22 existing field-studies based on the information available in the publications against nine specific criteria. The evaluation criteria should serve to interpret existing results and guidelines for future test study design. The 22 evaluated real-life test were evaluated according to the scheme below:[1]

Photocatalytic associations in Europe

The Photocatalytic Associations in Europe have played a significant role in establishing and promoting standard test protocols. The very basic of promoting a photocatalytic product is to use a catalytic product of sufficient activity. Hence, the product's activity should initially be tested and reported to ensure that the product is functioning (e.g., on a laboratory scale). The most widely recognized test procedure applied to determine the activity of NO_X-removing building materials using photocatalysis is the ISO 22197-1 standard test.

Spanish/Iberian Photocatalytic Association

photocatalytic white book

Fig. 17-7 The Photocatalytic white book by the Iberian photocatalytic association.

The leading Photocatalytic Associations for promoting the photocatalytic technology in Europe is the Spanish/Iberian Photocatalytic Association. In 2020, the Iberian Photocatalytic Association published a photocatalytic "white book" (Fig. 17-7). The leading commercial companies promoting photocatalytic products in Spain, as well as public researchers and public servants, also contributed to the white book.

The Iberian Photocatalytic Association also played an important role in making photocatalysis a technology used in public tenders. In Madrid, the building companies get a premium when offering city infrastructure comprising photocatalytic technology. The city of Madrid has been active within photocatalysis for many years, and the city of Barcelona and Malaga are the next in line to join the pro-active approach laid out by Madrid.

A major European funded Life project, LIFE-PHOTOSCALING (scan the QR code on the right) was conducted in the City of Madrid with support from the Iberian Photocatalytic Association. In this study, two key environmental problems were addressed:

1: Insufficient air quality in the large cities in Europe
2: The emerging problem of exposure to nanoparticles.

The project kicked off in October 2014 and ended June 30, 2019. The project's main objective was to scale up the photocatalytic technology in urban agglomerations from laboratory measurements to application in cities. The photocatalytic effect was investigated on both lab scale, pilot scale, and on two real-life projects in Madrid. In the Life-Photoscaling project, it was concluded that implementing photocatalytic technology in the City of Madrid is expected to reach about a 5% reduction of NO_X in five years and 15% in ten years.

A survey in the LIFE-PHOTOSCALING project also revealed that, among the interviewed people (916 people) 36% of the people had heard about photocatalysis, which is a remarkably high number compared with people across Europe.

Even though there is no photocatalytic association in Denmark, we see cities in Denmark actively deciding to use photocatalytic products. The

City of Frederiksberg, a part of Copenhagen, decided on April 30, 2018, to actively use photocatalytic pavement whenever new pavement was to be installed in the City (scan the QR code on the right). The Environmental Civil Service Office in Frederiksberg had, during a six-month period, investigated the photocatalytic technology and concluded, that it is proven that photocatalysis works and that it can be considered a useful tool for removing NO_X in high populated cities. Furthermore, the City of Frederiksberg concluded that adding photocatalyst in the concrete pavement would only increase the installation cost by 1-4 % per m^2.

In Summer 2020, Frederiksberg City included asphalt roads for testing photocatalytic technology, and the city has added photocatalysis into the 2030 Air Quality Plan.

Future outlook for photocatalysis in Europe

Looking back from when photocatalysis was initiated in Europe, a few crucial developments have altered the way we consider photocatalytic technology today. When photocatalysis was introduced in Europe, the technology was expensive. The price for a product which included the technology often cost twice the amount of a product without the technology. However, the development of going from the principle of using TiO_2 powders and introducing it in the entire matrix of the product to developing transparent water-based TiO_2 dispersions that could be introduced only on the surface of the product made an important change in the cost-benefit proposal. Not only was the price significantly lowered, but the performance was also improved considerably.

We are now in the 3 rd generation of photocatalytic materials, with an attractive cost-benefit offering, supported by substantial scientific documentation and documentation of the effect in real-life.

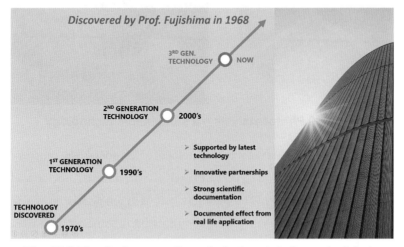

Fig. 17-8 The 3rd generation of photocatalytic materials in Europe.

We believe that the next five years of development for photocatalysis in Europe will be centered around integrating photocatalytic technology features and presenting it as a sustainable technology. Completing the Sustainable Circle for photocatalytic technology will be the next major improvement for photocatalysis in Europe. When the photocatalytic technology is considered and accepted in a sustainable scheme, we foresee another big increase in photocatalytic products sold, which also was the case when the documentation and knowledge about the reduction of NO_X was completed. We also expect an increasing awareness of photocatalytic technology in Europe, intending to reach the level of awareness observed in Spain in the remaining part of Europe.

Fig. 17-9 Photocatalysis – a sustainable technology

Talking about a sustainable technology or business, we generally address three main categories: Planet (Environment), Social and Economic sustainability. The photocatalytic technology has proven to have an impact on the environment with its NO_X-reducing properties. We believe that we will see photocatalysis as a climate technology in the future, where we look at the CO_2 footprint and reduction of Green-House Gasses (GHG). The CO_2 footprint from photocatalysis is known to have the lowest CO_2 footprint comparing NO_X-abatement technologies. It uses no additional energy in the use phase except from the sun's energy. All other competitive technology uses energy, for example, in the form of heat or higher fuel consumption.

The photocatalytic technology also keeps surfaces cleaner over time. This influences the cost of keeping the surfaces clean, the CO_2 footprint, the aesthetic appearance of the surfaces, and the thereby related social improvement. In 2008 a Life Cycle Analysis (LCA) based on the MEKA method showed that a self-cleaning window lowered the CO_2 footprint compared to regular window cleaning by 90%. A similar reduction will be obtainable with self-cleaning surfaces on other building materials.

The development of a documentation package for the self-cleaning effect for building materials comparable to what is available for NO_X reduction will drive the self-cleaning property in Europe in the coming years.

A new field for photocatalytic technology will be photocatalytic products to improve and enhance the products' performance to reduce the "heat-island" effect. Lighter colored surfaces better reflect solar radiation. Usually, the albedo of surfaces decreases over time due to dirt, tire wear, and algae growth. Photocatalytic products have the potential to reduce this decay by keeping the surface cleaner over time.

The second and third categories related to sustainable technology are the social and economic benefits. Today, it is possible to quantify the effect of the photocatalytic technology related to the NO_X-removal property and thereby related reduced societal health cost. Introducing

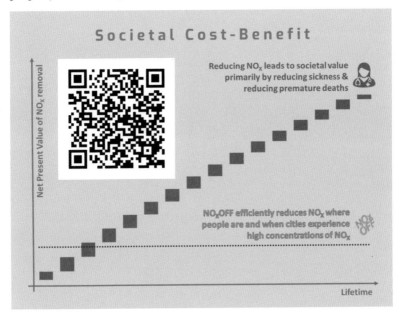

Fig. 17-10 Estimates of NO_X removal costs based on real-world environmental studies in Roskilde.

the Net Present Value of NO_x removal, a pay-back time will be reached within 2-3 years (Fig. 17-10).

References

1. P. D. Pedersen, N. Lock and H. Jensen, "Removing NO_x Pollution by Photocatalytic Building Materials in Real-Life: Evaluation of Existing Field Studies" *J. Photocatal.* 2021, 2, 1-13, https://www.eurekaselect.com/192128/article.

2. Environmental Industries Commission (EIC), "Towards Purer Air: A review of the latest evidence of the effectiveness of photocatalytic materials and treatments in tackling local air pollution."

(Henrik Jensen, Ph.D., CTO and Co-Founder at Photocat A/S, www. photocat.net)

[On behalf of the authors]

Akira Fujishima

Professor Emeritus, Tokyo University of Science
Special Professor Emeritus, The University of Tokyo
Born in Tokyo in 1942. Graduated from the Yokohama National University School of Engineering in 1966. Completed a doctoral program at The University of Tokyo Graduate School of Engineering and obtained a Ph.D. in engineering in 1971. He became a lecturer at the Kanagawa University Faculty of Engineering in the same year and became a lecturer at the University of Tokyo in 1975. From 1976 to 1977, he was a postdoctoral researcher at the University of Texas at Austin. He became Associate Professor, Faculty of Engineering, University of Tokyo in 1978 and became a Professor at the University of Tokyo in 1986. He became President of the Kanagawa Academy of Science and Technology in 2003. He was appointed as Professor Emeritus of the University of Tokyo in 2003 and later became Special Professor Emeritus of the University of Tokyo in 2005. In 2010, he became President of the Tokyo University of Science and was appointed Professor Emeritus at the Tokyo University of Science in 2018.
He is currently Director of the Photocatalysis International Research Center at Tokyo University of Science, Chairman of Tokyo Ohka Foundation for The Promotion of Science and Technology, Chairman of the Photofunctional Materials Research Association, Professor Emeritus of Peking University, Professor Emeritus of Jilin University, Professor Emeritus of Shanghai Jiao Tong University, Professor Emeritus of the University of the Chinese Academy of Sciences, Member of the European Academy of Sciences, and Foreign Member of the Chinese Academy of Engineering.
He has served as Chairman of the Electrochemical Society of Japan, Chairman of the Chemical Society of Japan, and a member of the Science Council of Japan.

【Awards】 Order of Cultural Merit in 2017, Thomson Reuters Citation Laureates in 2012, The Luigi Galvani Medal in 2011, Person of Cultural Merit in 2010, Kanagawa Culture Award in 2006, Imperial Invention Award in 2006, Japan International Prize in 2004, Japan Academy Prize in 2004, The Prime Minister's Award for Industry-Academia-Government Collaboration in 2004, Purple Ribbon Medal (Shijuhosho) in2003, 1st Gerischer Award in 2003, Chemical Society of Japan Award in 2000, Inoue Harushige Prize in 1998, Asahi Prize in 1983, etc.

He is currently an Honorary Citizen of Kawasaki City and Toyota City. He is the author of 950 original papers (in English only), about 100 books (including contributed works), about 500 review articles and commentaries, and 310 patents.

[coauthor]

Tsuyoshi Ochiai 【Chapters 3・6・8・9】

Senior Researcher
Local Independent Administrative Agency Kanagawa Institute of Industrial Science and Technology (KISTEC), Kawasaki Technical Support Department, Materials Analysis Group
Adjunct Lecturer, Hosei University

Kengo Hamada 【Chapters 4・5・14】

Researcher
Local Independent Administrative Agency Kanagawa Institute of Industrial Science and Technology (KISTEC), Kawasaki Technical Support Department, Materials Analysis Group

Donald Alexander Tryk 【English Translation】

Professor

University of Yamanashi
Fuel Cell Nanomaterials Center

Chiaki Terashima 【Chapter 12】

Professor
Tokyo University of Science, Research Institute for Science and Technology
Photocatalysis International Research Center
Research Center for Space Colony

Norihiro Suzuki 【Chapters 4・5・7・11】

Junior Associate Professor
Tokyo University of Science, Research Institute for Science and Technology
Photocatalysis International Research Center
Research Center for Space Colony
(Present Affiliation: Photocatalysis International Unit, Research Center for Space System
Innovation)

Katsunori Tsunoda 【Chapter 9】

Project Manager
Tokyo University of Science, Noda Research Support Section

Hitoshi Ishiguro 【Chapter 10】

Vice-Leader
Local Independent Administrative Agency Kanagawa Institute of Industrial Science and
Technology (KISTEC)
Research and Development Department
Photocatalyst Group

Jinfang Zhi 【Chapter 15】

Professor
Chinese Academy of Sciences
Technical Institute of Physics and Chemistry

Jong-ho Kim 【Chapter 16】

Professor
Chonnam National University
Chairman of Photo & Environmental Technology Co., Ltd. (South Korea)

František Peterka, Ph.D. 【Chapter 17】

Co-founder of CEN/TC 386, the European Committee of Standardization on Photocatalysts
Co-founder of Czech Association for Applied Photocatalysis (CAAP)
Special Member of the Photocatalysis Industry Association of Japan (PIAJ)

Henrik Jensen, Ph.D. 【Chapter 17】

Chief Technology Officer (CTO) and Co-Founder, Photocat A/S, Denmark, www.photocat.net

Photocatalyst Museum Office (Tomoko Aoki・Momoko Tsurumi) 【Chapter 13】

Local Independent Administrative Agency Kanagawa Institute of Industrial Science and
Technology (KISTEC)
Research and Development Department, Research Support Section, Research Support
Group

Akira Fujishima and Tsuyoshi Ochiai appeared on following YouTube program on 2021.11.12

The latest information in simple terms
Photocatalysis Experimental Methods
〈英語版〉最新情報をやさしく解説　光触媒実験法

2021 年 12 月 24 日　第 1 刷発行

Co-writers: Akira Fujishima, Tsuyoshi Ochiai, Kengo Hamada, Donald Alexander Tryk, Chiaki Terashima, Norihiro Suzuki, Katsunori Tsunoda, Hitoshi Ishiguro, Jinfang Zhi, Jong-ho Kim, František Peterka, Henrik Jensen and Photocatalyst Museum (Tomoko Aoki and Momoko Tsurumi)
Editors: Yoshinobu Kitano and Kiyoshi Matsushita
Illustrators: Norihiro Suzuki and Kengo Hamada
Publisher: Yoshinobu Kitano
Published by Kitano-Shoten Publishing Co.,Ltd.
KITANO Building 3F, 1-18-7 Kashimada, Saiwai-ku, Kawasaki-shi, Kanagawa 212-0058
http://www.kitanobook.co.jp
Tel: +81-44-511-5491
info@kitanobook.co.jp
Printer: Taihei Printing Co. Ltd.

©2021 Akira Fujishima
ISBN 978-4-904733-08-0
If there are any pages out of order or missing in your copy, we will replace it. Please note that we cannot replace books purchased at used bookstores.
No reproduction or republication without permission.
Printed in Japan

著　者 ——— 藤嶋 昭
落合 剛・濱田健吾・Donald Alexander Tryk
寺島千晶・鈴木孝宗・角田勝則・石黒 斉・
只 金芳・金 鍾鎬・František Peterka・
Henrik Jensen・光触媒ミュージアム事務局
（青木智子・鶴見桃子）
編　集 ——— 北野嘉信・松下清
図版作成協力 — 鈴木孝宗・濱田健吾
発行人 ——— 北野嘉信
発行所 ——— 株式会社 北野書店
〒 212-0058 川崎市幸区鹿島田 1-18-7
KITANOビル 3 F
http://www.kitanobook.co.jp
電話／044-511-5491
印　刷 ——— 株式会社 太平印刷社

©2021 Akira Fujishima
ISBN 978-4-904733-08-0
落丁・乱丁の場合はお手数ですが小社出版部宛にお送りください。送料小社負担にてお取替えいたします。但し、古書店で購入されたものについてはお取替えできません。
無断転載・複製を禁ず
Printed in Japan